Beyond empiricism

Monographs in Social Theory

Editor: Arthur Brittan, *University of York*

Titles in the Series

Barry Barnes   *Scientific knowledge and sociological theory*

Zygmunt Bauman   *Culture as praxis*

Keith Dixon   *Sociological theory*

Keith Dixon   *The sociology of belief*

Tom W. Goff   *Marx and Mead*

Bernard Meltzer et al.   *Symbolic Interactionism*

Anthony D. Smith   *The concept of social change*

A catalogue of books in other series of Social Science books published by Routledge & Kegan Paul will be found at the end of this volume.

Andrew Tudor

Beyond empiricism
Philosophy of science in sociology

Routledge & Kegan Paul
London, Boston, Melbourne and Henley

First published in 1982
by Routledge & Kegan Paul Ltd
39 Store Street, London, WC1E 7DD,
9 Park Street, Boston, Mass. 02108, USA,
296 Beaconsfield Parade, Middle Park,
Melbourne, 3206 Australia
and Broadway House, Newtown Road,
Henley-on-Thames, Oxon RG9 1EN
Printed in Great Britain by Redwood Burn Ltd,
Trowbridge, Wiltshire.
© Andrew Tudor 1982
No part of this book may be reproduced in
any form without permission from
the publisher, except for the quotation of brief
passages in criticism

Library of Congress Cataloging in Publication Data

Tudor, Andrew, 1942–
Beyond empiricism.

(Monographs in social theory)
Includes bibliographical references and
index.
1. Sociology – Methodology. 2. Science –
Philosophy.   I. Title.   II. Series.
HM51.T77   301'.01'8   81–17916
ISBN 0–7100–0925–9   AACR2

for Jamie with love

As regards those who adopt a *scientific* method, they have the choice of proceeding either *dogmatically* or *sceptically;* but in any case they are under obligation to proceed *systematically*.
Immanuel Kant, 'Critique of Pure Reason'

Contents

Preface — ix

1 The philosopher's stone — 1

2 Practical epistemology — 14
   A common-sense epistemology
   The scientific canopy
   The web of belief

3 Theories and things — 45
   The received view: basic character
   The received view: problem areas

4 Elements of theory — 73
   Languages, models and sentences
   The rules of inquiry

5 On explanation — 91
   Explanation and covering laws
   Explanation by model
   Explaining and interpreting

6 On demonstration   118
   Conjectures and refutations
   Conventions, basic statements, and corroboration
   From corroboration to incommensurability
   On demonstration

7 Methodological diversity in sociology   157
   Dimensions of variation
   Complementarity and conflict
   Sociology, naturalism, and the philosophy of science

   Notes   183

   Name index   205

   Subject index   209

Preface

As someone who was an undergraduate in the early 1960s I am, I suppose, a member of the last generation of sociology students for whom the philosophy of science seemed centrally important. Much concerned with the methodological status of our chosen discipline, we looked to philosophers of science for clarification and guidance. I'm not sure now how much of that we actually received, but it is undeniable that the image of science that we encountered influenced profoundly our conceptions of the sociological enterprise. When, a very few years later, the tenuous web of methodological consensus was finally blown away, so, too, was sociology's habitual reliance on terms begged, borrowed, or stolen from the philosophy of science.

Or so it seemed. In fact it wasn't long before the old topics re-emerged, though this time as part of a campaign against such sins as 'positivism' and 'empiricism'. Faced with this flood of seemingly philosophically informed rhetoric, in 1973 I decided to write a short book on positivism and sociology – as much as anything an attempt to clarify for myself some of the issues embedded in current arguments. The present book is a great-grandchild of that project, its features much changed by the fact that philosophy of science, when I re-encountered it, was itself almost unrecognisable. The dissent which was only beginning in the early 1960s had turned into open war by the end of the decade. As I set about 'catching up' I found myself increasingly uncertain about my ability to grapple with the problems generated by this disorder, and the work proceeded in fits and starts. More than once I abandoned the manuscript, drifting away toward more journalistic and less technical pursuits.

I mention all this not to excuse in advance any incoherence evident in what follows, but rather to suggest the tentative nature of this study. I still believe that philosophy of science has something to offer to sociology, but in a spirit of co-operative interchange of ideas rather than one of benevolent guidance. Accordingly this book seeks to explore some features of contemporary philosophy of science from the point of view of their utility for sociology's self understanding. That is inevitably incomplete and one-sided - I have no pretences to writing a definitive text-book - but not, I hope, so much so as to be merely another shot in sociology's long-running methodological war. That is not how I intend it.

Over the years that I have been writing the book I have learned from many people, some of whom have been kind enough even to read and comment upon the manuscript. I cannot list all to whom I am grateful here, but I would like to thank Jerry Booth, Ruth Bush, Bob Coles, Alan Dawe and Mary Maynard who have, wittingly or unwittingly, contributed to this study. They will, I am sure, disagree with much of what I say, but, then, that is as it should be. I am also indebted to Dorothy Lane for her speedy and good-humoured translation of my manuscript into type, and I would particularly like to thank Arthur Brittan, who has been both friend and editor, and who encouraged me to finish the work at a time when I had given it up. I am, of course, responsible for any errors and infelicities contained herein, but he is responsible for making them publicly available.

Andrew Tudor
November 1980

chapter 1

The philosopher's stone

> The distinction of science from all the philosophical disciplines is vital. It will turn out to be so at every stage of the ensuing study. But this is <u>not</u> to be taken to mean that the two kinds of discipline are without significant mutual interrelations and that each can afford to ignore the other.
> Talcott Parsons, 'The Structure of Social Action'

Sociology is a mass of bad habits, not the least of which is the sociologist's own willingness to chronicle them as such. But, that paradox apart, it is clear that one of the worse indiscretions has been a thoroughly cavalier attitude to philosophy. Having rightly concluded that philosophy was of some importance to the sociological enterprise, sociologists (and I am one) have used that discipline much as the military might use a guided missile. Safely fired in the conviction that it will seek and destroy, the soldier need know little of the missile's true workings and consequences. Likewise for the sociologist. Recognising the incipient power of labels borrowed from philosophy, sociologists have strewn them about with little regard to their detailed significance. Indeed, if armies were so irresponsible (and they may yet be) I should not be writing, nor you reading, this essay. We would have long since vanished in drifting clouds of nuclear fallout. And it is because I fear a destructive philosophical fallout in sociology that I have embarked on this project.

Not that there is any shortage of discourse on the manifold relations between the two disciplines; far from it. But many of those discussions are basically unhelpful, sometimes even misleading and problematic. Those produced by philosophers are, naturally enough, philosophically sophisticated, if indefatigably ignorant of the constraints of sociological practice. They tend toward philosophical imperialism. Those produced by sociologists, on the other hand, use philosophy quite unashamedly for their own ends. (1) I have no objection to that strategy; after all, knowledge is there to be used. But promiscuously consorting with philosophical labels - increasingly common in sociology - confuses the issues rather than clarifying them.

For reasons which will emerge I am particularly interested in the relations between sociology and one specific area of philosophy: the philosophy of science. There is a well-established tradition in sociological methodology that treats the philosophy of science as a kind of philosopher's stone: a source of magical instruction on how science ought to be done. Only grasp that stone with both hands and the base metal of sociology will be transmuted into the gold of scientific maturity. This view is now rightly in somewhat bad odour, both for sociologists who do not see the legitimation of their discipline as something to be found elsewhere, and for philosophers who see the philosophy of science less normatively than once they did. Nowadays the point of epistemology in general, and philosophy of science in particular, is less the <u>validation</u> of the methods of empirical science, and more <u>understanding</u> the links between empirical world and scientific knowledge. (2) Because of this changing emphasis on the part of philosophers it is a pity that many sociologists have reacted away from philosophy of science concerns. I should like to argue for their reinstatement, but as part of a process of understanding and not as a source of prescription. Philosophy of science cannot be a philosopher's stone as it was once thought to be, but it can serve as something of a crystal ball to aid our self-reflection.

The philosophy of science that concerns me centrally here is the dominant tradition, rooted firmly in the Vienna School and logical empiricism. As I shall suggest in chapter 3, it has moved a good way from these beginnings, progressively liberalising the very strong assumptions embedded in the original philosophy. Unfortunately, however, such nuances have often been lost on sociology, and I am sure that my interest in this tradition will be dismissed as <u>positivist</u>, that currently fashionable catch-all criticism. Not that such an accusation says very much: it is impossible to pin down any stable source for sociological usage of the term, any coherent account of the meaning of the doctrine. It is one measure of this confusion that I have even heard a graduate sociologist classify Schutz's sociology as a 'positivist' development of Weber, with every sign that he thought this to be an important insight. In fac', most sociological users of the expression try to invoke a hazy sociological tradition as well as some presumptively concomitant philosophical perspective. As far as the sociological element goes, the classification can and does include almost any perspective up for castigation. In the sociologist's account of 'philosophical' positivism, however, the range is more limited, generally emphasising and subsuming two sorts of positions within the one conception. The standard labels here are 'naturalism' and 'empiricism' and I shall return to them shortly.

For the moment let me note a curiosity about sociological applications of the positivist stigma: if anything they under-emphasise the one characteristic truly constitutive of positiv*ism* - the claim that the world can only be known scientifically, and that we can thus only have knowledge of that ontological dimension amenable to such methods. In short, that there is a single, observable, factual reality, and all else fails the test of knowledge. (3) If this is the kernel of positiv*ism*, and I think it is, then it provokes my first disavowal. I do not believe that the knowable world can be sensibly thus limited, nor that our understanding should have such strictures placed upon it. If positivism rests on this uni-dimensional ontology then it is not for me; more important, nor has it been of any real significance in the development of contemporary sociology. Few, if any, sociologists have been short-sighted enough to advance such a doctrine. Even psychological behaviourism, which approaches it most closely, is far from uncontaminated with extra-positivist traits.

So, any residuum left to this study that might still be labelled 'positivist' has nothing to do with such ontological myopia. What I do echo from 'logical positivism' is a belief that there are crucial issues surrounding the question of empirical demonstrability ('empiricism') and a belief that there is a single _general_ mode of empirical inquiry within

which we find inevitable detailed variation ('naturalism').
If this is <u>positivism</u>, then so be it. I do also hold a 'positive' view – though now the play is on words – in that I believe that sociology can and should make a positive contribution to the worlds on which it is parasitic, rather than 'progressing' toward a professional and academic reflex of whatever happens to be the current status quo. I am in good company here, at least, though perhaps I shall lose these briefly gained friends for fear of my other symptoms. Yet I do think the two must go together: a concern with word and object, theory and thing, marks the road we must travel to arrive at an acceptable and realisable positive commitment. Sociological 'demystification' depends on our awareness of the grounds on which we believe that sociology can indeed render up a demystified version. (4) And in seeking that understanding we are thrown in at the philosophical, epistemological, and methodological deep end.

Thus I am returned to the two sub-views that I suggested underlie much guerilla use of the positivism gambit. Let me first consider the less problematic of the two: 'naturalism'. The philosophy of science mainstream was historically much concerned with the 'unity' of the sciences, a concern which also appears in the more directly sociological contributions. Though it was important, in this context I shall take that debate as given. But in the general epistemological discussions of chapter 2 I shall imply, at some level, that all empirical disciplines share a basic structure of inquiry. In this respect the 'naturalism' propounded here is not that which Morganbesser would ascribe to a 'canonical social scientific naturalist', one who believes in the <u>reducibility</u> of

the social sciences to the natural. (5) Rather, it is the
<u>methodological naturalism</u> that he ascribes to Nagel, different versions of which may be found throughout the philosophy of science. Given all this, though, it still remains to identify the structure of inquiry in question, and the level at which it can sensibly be said to be common to empirical analysis. Much of this book is concerned with different aspects of the first problem. On the second I should note two well-established families of arguments, even though I have no intention of pursuing them here.

The first is that classically raised in the methodological writings of Max Weber and in the German intellectual context in which he was working. At its simplest the argument is that different subject matters require different methods, a plausible proposition traditionally incarnated in the historic distinction between the 'natural' and 'cultural' sciences. Inevitably the argument hinges on what is meant by 'method'. If the reference is very specific, then it is virtually a truism that different subject matters demand different methods. If it is not this specific then the issues become very complex; hardly amenable to brief treatment. I must ask to be allowed to beg this question. The second family of arguments is more practically oriented. They make the claim that 'scientific method' (if there is such a dogma) demonstrably fails when transferred wholesale into sociology. This is quite true; the doctrines espoused by unwary philosophers of science do not work in sociology. Significantly enough, they do not work in science either, which is not practised as many pristine philosophies of science have led us to believe. (6) This is part of the reason for the changes in the discipline that I have already mentioned. At best, however, such philosophies have provided us with 'rational reconstructions' of certain aspects of the process of inquiry, partial versions of what passes as science. Sometimes the philosophers themselves have been guilty of reifying their own accounts, though it has been more common to find sociological borrowers misunderstanding, misinterpreting, or misusing them. Provided we recognise their inevitable limitations, rational reconstructions can be quite useful. I shall return to the whole question in chapter 2.

In the present state of the game many sociologists would claim that the 'science or not' argument is dead, resolved in one direction or the other. And, as it was traditionally posed, if not actually buried yet, the debate is indeed sterile. So why should I even mention it here? For one thing, of course, it was very much involved in sociology's traditional misconstruction of philosophy of science as a philosopher's stone. For another it is sometimes argued that precisely because the 'science or not' debate is now irrelevant we need no longer concern ourselves with the sorts of issues it historically served to raise. But the opposite case is just as plausible. Because important issues of methodology and epistemology were historically raised in this particular form, we can easily be deluded into thinking that the issues are as sterile as the debate has become. And that is far from true. As practitioners of what is, as yet, a somewhat incoherent discipline, we must still be crucially concerned with the epistemology of sociology and, ultimately, with epistemology in general. The virtue of methodological naturalism lies in its insistence that we identify a general structure of inquiry in some detail. I believe this to be a viable and profitable enterprise, but not one which can be properly justified in advance of the enterprise itself. And that is part of the point of this book. So, if 'naturalism' is positivistic I must plead guilty as charged. But in view of the stereotypes so common in sociology I should perhaps stress that this 'naturalism' is very much a revisionist view. I am as far from believing that the social sciences should 'ape' the natural as human beings are from the apes themselves. And if that metaphor seems more complex than it might at first appear, that is just as well. So is its subject.

Rather similar considerations apply to the second sub-view I identified, though I think the issues here are probably wider and more complex. Certainly the questions demand more inclusive answers. But once more there is the problem of pernicious sociological stereotyping: 'empiricism', whatever it might mean, is an over common pejorative in sociological discourse. Once only applied to sociological researchers of the persuasions Alison Lurie delightfully labels 'nuts and sluts' and 'the numbers game', the term has increasingly found application to almost anyone with pretensions to make general assertions about the empirical world. (7) The Willers have even seen fit to develop a cogent critique of much sociology as 'systematic empiricism', within which ambit they include all sociologies that focus primarily on establishing empirical generalisations. Many of their views (in this particular work and elsewhere) are persuasive, and their contribution appears to have been sadly ignored by the sociologists at whom it is directed. (8)

But because they use 'empiricism' in this generalised sense they are obliged to omit certain nuances which are crucial to any closer discussion. They, and other sociologists who have written less persuasively on this subject, take for granted a single 'empiricist' tradition, or, at least, a set of philosophies bearing a very clear family resemblance. Locke, Hume, and Mill figure prominently in the classical background, though discussion of modern developments is greatly attenuated. It is rare, for instance, to find much consideration given to post-Carnap empiricist philosophy or the various 'revised' empiricisms of recent years.

The omission is significant, for 'empiricism' may also be seen as a partial perspective, reified and mobilised in sociological discussion as either slander or received wisdom. But 'empiricism' has changed, especially with the collapse of the view that we must seek absolute justifications for our methodologies, absolute 'truth' in our researches. I am very much concerned with 'empiricist' issues, though if I espouse any empiricism it would be a revisionist version of revisionist empiricism. In Scheffler's term, no more than a 'modest empiricism', and certainly a sociologist's vulgarisation and misuse of views advanced in modern philosophy of science. (9) The image of philosophical empiricism most commonly found in sociology ignores such recent developments, sticking to the much more rigid ideas of the classical formulation. If only because of this gap, modern philosophy of science and epistemology must have some clarification to offer to the sociological reader.

So, once more I must ask the reader tolerantly to look behind the label and accept that the 'empiricist' element in the umbrella concept 'positivism' is not quite the straightforward sin that its classical expression suggests. It could be asserted (though not too persuasively) that this book is positivist, naturalist, empiricist, all three; however, it is time to consider the whats and whys of such commitments rather than dismissing them in ill-assorted slogans. To underline that need perhaps I should invoke yet another <u>ism</u> imbuing this book, often wrongly imagined to be inconsistent with the position I have been sketching thus far. I have in mind <u>relativism</u>, the ghost at the feast of so many modern disciplines. Long a familiar issue for sociology, recent philosophy of science has spread the word far and wide through the influential work of Thomas Kuhn. Questions he has served to raise - though he has many forebears and notable contemporaries - run through this book, just as they do through contemporary philosophy of science. But I cannot begin to decipher the meanings of <u>relativism</u> at this stage without anticipating the detailed discussions of the rest of this study, an inability which should serve to remind us yet again that labels are labels and arguments are arguments. I hesitate to suggest that the twain might never meet, but I fear that they have rarely done so in the sociophilosophical literature or, indeed, in this introductory chapter.

With what, then, am I left? At the beginning of this chapter I suggested that philosophy had been unhappily related to sociology; partly because of philosophical imperialism, occasionally from sociological imperialism, but mostly because sociologists have combined hyperconsciousness about philosophical issues with simple-minded views of philosophical schools. Hence the uncritical reliance on received labels, not simply <u>isms</u>, but also in the use of such terms as 'correspondence' and 'coherence' theories of truth, 'deductive' theoretical structure, 'falsification', and the like. I am not a philosopher, so I, too, suffer from these ills. For this reason alone (though there are many others) I must limit my task. Many books on the philosophy of social science try to run the whole gamut of issues, and although it is clear that arguments about one aspect of philosophy connect up with arguments about another, there is something to be gained from a more restricted area of discourse. Hence my focus on the historically important philosophy of science, on problems of understanding the structure of sociological inquiry, and hence, too, my neglect of the equally significant philosophy of action tradition. Philosophy has much to say to sociologists about the nature of human action, communication and thought; the absence of such topics here reflects only a convenient division of labour.

In a word, then, my interest is <u>formal</u>. This does not imply that I think sociology can or even should be 'value-free', that there are not important philosophical issues about the <u>substance</u> of sociological inquiry, or that all our problems will miraculously dissolve if we can make a methodological settlement. It simply means that I think formal issues are important issues and that they run the risk of neglect because of the way they (and the philosophy of science) have been traditionally invoked in sociology. The time is ripe for renewed discussion. For any professional philosophers who might happen upon this volume, however, let me say that my haphazard philosophical education, and the present state

of sociology, make the book philosophically unsatisfactory.
And, if I use philosophy here in the 'underlabourer' mode,
it is not necessarily because I agree with that as a limitation but because I am writing for sociology and sociologists.

The form in which subsequent discussion is cast becomes
a little complex at times, and the overall pattern is hardly
linear. Rather than try to outline the argument here (I
shall try to do so at the end of each chapter) I will simply
indicate the overall order of topics. It follows fairly naturally from the assumptions I have been setting out that I
must first discuss knowledge in general. Accordingly,
chapter 2 is concerned with 'practical epistemology' and
develops a highly programmatic account of empirical inquiry.
Chapter 3 then digresses into one particular specification of
<u>scientific</u> inquiry, historically very influential in the philosophy of science and imported in odd ways into sociology.
This revolves around the once orthodox conception of
'theory' in science, dubbed, by Putnam, the 'received view'.
Although apparently something of a digression, this discussion provides background essential for understanding more
recent philosophy of science and hence for the second half of
this book. Chapters 4, 5, and 6 return to the framework
advanced in chapter 2, 'filling in' some of the blanks in that
schematic summary. Chapter 4 focuses attention on the
components of theory, chapter 5 on questions about 'explanation', and chapter 6 on 'demonstration'. Needless to say,
that discussion does not exhaust even the issues raised in
the first half of this study; I do not intend to be comprehensive. Finally, in chapter 7, I shall return to sociology and
to the sort of understanding philosophy of science might offer

to us. Not a philosopher's stone certainly, but a tool - useful or useless like any other.

chapter 2

Practical epistemology

> Science is self-conscious common sense. W.V. Quine, 'Word and Object'

Twentieth-century sociologists have often doubted the use of discussions like this one. And now that we find ourselves in a situation where sociology has moved from treating epistemology as nothing at all to recoiling in horror before the full force of the relativistic whirlwind, there may be even more reason for questioning the value of yet further epistemological discussion. Both views have something to be said in their favour. The former because it is often difficult to see how abstract discussion of the 'nature of knowledge' can be useful to a practising sociologist; this is the province of philosophy, and philosophy has different aims. The latter because modern 'sociological' epistemologies seemingly lead to a vicious regress: how can I know that I know that I know...? How indeed?

There are senses in which I subscribe to both positions. Much extant philosophical discussion of knowledge is not as such germane to sociological concerns, for the gap between philosophical theorising and sociological practice is too wide to be usefully and quickly bridged. Even that epistemological specialism the 'philosophy of science' has been so dominated by the natural sciences, indeed by physics, as to raise serious problems for itself in application to sociology. It can be argued that the philosophy of science might benefit from considering the problems set by social science. But it is equally arguable that modern sociology's obsession with the epistemological barriers to its own disciplinary 'progress' is tragically negative. It is one thing to recognise that there are unavoidable limitations on our truth-claims; it is quite another to be paralysed into dealing only with trivial issues because the complex ones raise too many philosophical problems. If sociology cannot bring itself to make serious, responsible, and committed statements about the social world then there is little point to sociology.

However, the degree to which sociology raises special epistemological problems is not a question I shall take up directly. Any comfort I might have to offer to the epistemologically distressed will emerge in the normal course of discussion and not as a special topic. But I do believe that a philosophically and sociologically sensitive epistemology is an important part of the sociological enterprise, and so, to that degree, I would wish to bridge the gap between the two. What is more, if I can be allowed some momentary latitude on questions concerning the social character of knowledge, I think I can justify importing more strictly philosophical referents into this discussion. Indeed, I could hardly avoid them. But this means that, however sketchily, I must begin by clarifying the role such materials might reasonably play: how they stand in relation to the bundle of practices called sociology.

It is obvious that this has frequently been a subject of debate in relation to other disciplines: most notably, the relation of science to the philosophy of science. I have already made some commitment here for I have accepted the usefulness of philosophy in the 'underlabourer' role, though without the stronger claim that this exhausts its use. And this commitment is already on the road away from philosophy conceived as a source of methodological prescription, a body of discourse which tells us how we should set about doing science. With the possible exception of some of the purer Popperians, who appear to preserve some sense of prescribing the 'oughts' of rational inquiry (some of that work will be discussed in chapter 6), this shift is in line with modern trends in the philosophy of science. (1) Here the term 'rational reconstruction' is informative: a philosophical approach devoted to analysing out an ordered version of existent scientific practices. (2) Quite how far we may legitimately take this ordering process is a subject for debate. Goodman suggests, for example:

> Philosophy, to my way of thinking, has rather the function of explicating scientific – and everyday – language than of depicting scientific or everyday procedure. While explication must respect the presystematic application of terms, it need not reflect the manner or order of their presystematic adoption; rather it must seek maximum coherence and articulation (3)

There is a lot to be said for this view of explication, for it is not completely divorced from scientific practice (though it is 'removed' in some degree), and it can serve to clarify the nature and status of the discipline under analysis. It is also consistent with methodological naturalism; if a general mode of empirical inquiry is to be isolated, rational reconstruction is likely to have more practical potential than a priori reasoning. It does at least touch on the processes of inquiry themselves.

Even so, rational reconstruction has two clear faults. First, like any theory, it provides us with a partial account. There is nothing necessarily wrong with that; no account of a phenomenon, complex or simple, can encompass every conceivable salient factor. The problem is not partiality itself, which is inevitable, but consistent failure to recognise that inevitability. There is always a danger that the rational reconstructor will delude himself or herself into believing that this version is the version; that whatever he or she has chosen to emphasise exhausts the subject. We can, and must, live with that danger, but only if we are always aware of it. This is the price of theorising. The second difficulty arises not from the 'reconstruction' but from the 'rational'. Naturally there is strong pressure on the rational reconstructor to select those aspects of scientific practice which best fit the rational ideal. But as such modern philosophers of science as Thomas Kuhn and Paul Feyerabend have been at pains to point out (and as the sociology of science has long recognised) there is a strong non-rational component in scientific inquiry. (4) All is not a neatly ordered bed of roses. Unfortunately, the version of scientific inquiry created by rational reconstruction is not always limited to simple 'ordering', or even to Goodman's 'explication'. It carries systematisation one step further by reconstruction in terms of <u>logical</u> order. One such argument, for example, notes that scientific explanation invariably seems to involve demonstrating that a particular phenomenon is a specifiable case of a more inclusive account; explanation involves subsuming the particular beneath the general. This link has been misleadingly reconstructed as a precise relation between a conditional sentence and the statement(s) which may be logically deduced from it. It is a case I shall discuss in detail later in this book. (5) For the present it is sufficient to note that there is a considerable difference between imposing a previously developed logical scheme and openly trying to systematise the process of inquiry: to indulge in, as Bunge has put it, 'systematic analysis'.

It is not an easy distinction, however, and it makes my chosen path that much more hazardous. I am advocating a 'systematic reconstruction' which heeds the strong case for social and cultural relativism. And just to complicate things more I want to use philosophical tools to dig out the elements of empirical inquiry, but without the prescriptive overtones so often associated with such discussions. So, while I do need schematic answers to classic questions like, 'what is knowledge?' or 'what is truth?', I do not wish to engage with the full philosophical debates which have surrounded these questions. It's not that anything I suggest here is meant to stand as <u>analytically</u> true. Since some part of the philosophy underlying this discussion doubts the feasibility of a clear distinction between <u>analytic</u> and <u>synthetic</u>, I could hardly make such claims. (6) Like all epistemology, this is a <u>theory</u> of empirical inquiry, conditioned by assumptions and available for 'test'. The fact that it is epistemology itself which gives precise meaning to 'theory' and 'test' complicates the issue but does not confound it totally. Epistemology tries to isolate and order the systems of knowledge in the light of which we are enabled to have knowledge about other aspects of our world; it is about 'knowing how' we know and not 'knowing that', to employ a somewhat bent version of Ryle's distinction. (7) And once it is deprived of the right to analytic truth and a priori knowledge, epistemology can also be seen to employ rules of evidence and inference in studying the operation of such rules in other disciplines. There can thus be an epistemology of epistemology, and so on into the regress.

It is a major failing of much epistemological discussion in modern sociology that it exaggerates the untoward consequences of this unavoidable circularity. This is one manifestation of the 'relativistic dilemma'. It becomes a problem because of a misguided, often covert, commitment to discovering absolute truth; to anyone thus committed the least smell of a regress must be anathema. But provided we recognise that we are not making any absolutist truth-claims the problem loses much of its bite. Epistemological discussion is then intended to increase awareness of what is involved in sociological inquiry so that there is some real possibility that we might recognise both the virtues and limitations of our claims to knowledge. I believe that the account I shall advance is little more than an explicit ordering of some of the crucial characteristics of sociological practice; an 'explication' in Goodman's term. Nowadays it is common for sociologists to appeal to their colleagues to be more self-reflexive. (8) What I am suggesting is reflexivity at its most general level: this is what I mean by 'practical epistemology'.

But how? Even given the goal, there are several strategies available. For instance, I could begin by considering the development of philosophical ideas about knowledge, and the relevance of such general theories of knowledge to particular disciplines like sociology. Certainly these theories figure prominently in some of the debates I shall have to invoke. But there are much better introductions available than I could ever hope to offer, even though their general point of view may not be the same as mine. (9) Instead of leaning directly on these proverbial angels, I shall, then, rush in, advancing an argument which leads from a simple-minded common-sense epistemology to a particular view of the basic components of all empirical inquiry. I do not intend the argument from 'common sense' to be a <u>justification</u> for what follows, though it has propaganda uses, and I do concur with the quotation from Quine that heads this chapter. (10) As much as anything, putting the argument in this form is intended as an aid to understanding my already inadequate prose.

A COMMON-SENSE EPISTEMOLOGY

In the course of our daily lives we are forever making claims to knowledge. Some of those claims may be intrinsically non-empirical (though they may have empirical consequences) - claims about ineffable, unknowable gods for example. Others are empirical in that they are seen to be about the 'real' world, about 'things that happen'. Needless to say this does not mean a world of simple observables. I can make empirical knowledge claims that someone is feeling pain or envy or that the economy is undergoing galloping inflation, just as I can claim that there is a tree at the bottom of my garden. Though there are variations in degrees of <u>abstraction</u> (from the various levels of what is ontologically claimed as the 'real world'), a large proportion of our knowledge-claims refer to 'true empirical knowledge'. Almost immediately, then, I find myself teetering on the edge of a philosophic abyss. I must continue to do so, for the abyss in question (the relation between belief and knowledge), though significant, can only distract from my main argument. It is sufficient to the occasion to note that embedded in what I have said thus far is a view of 'empirical' and 'scientific' knowledge as one form of belief. Nothing as simple as the equation 'knowledge equals true belief', but nothing so problematic as to make a precise distinction between the two necessary at this stage. However, the reader unschooled in past philosophical debates should be aware that the opposite point of view can be cogently held. (11) Much argument has been advanced about beliefs' 'psychological' character distinguishing it from knowledge, the sort of distinction which finally leads to concepts like Popper's 'third world' of 'objective knowledge', or at least to what Danto has called a 'de-psychologised' conception. (12) Such views are against the grain of <u>this</u> argument, and I can only suggest that interested readers pursue these fascinating philosophical intricacies in the literature.

So we do frequently make claims to having 'true empirical knowledge', though not everything we are prepared to label 'knowledge' is also necessarily true. That is especially obvious if we take temporal factors into account, for in 'truth', as in other matters, what is here today may be gone tomorrow. But always underlying such knowledge-claims is the common-sense context, and for us to accept knowledge in that context I think we must meet three general criteria.

1 Intersubjective legitimation

Knowledge-claims must be shared by a body of 'significant others' as well as simply advanced by an individual claimant. This rests on the observation that, for the most part, individuals will only accept something as 'true' insofar as it is socially, psychologically, and culturally acceptable to some group of people whose judgment is considered significant. The sociological, anthropological, and psychological literature furnishes many examples of this phenomenon. There are, for instance, those well-known experimental studies demonstrating the ways in which pre-arranged group 'error' can change individual answers to simple questions. When asked to count the number of times a light flashes, the subject follows the group decision at the expense of his or her own arithmetic. Or there is the sociologically commonplace observation of the importance of reference groups, opinion-leaders, and the rituals necessary for the affirmation of knowledge and belief in primitive and not-so-primitive societies. There are also 'deviant' sub-cultures – like the flat-earthers – who are able to make truth-claims in the face of majority disagreement precisely because they can form their own cohesive set of 'significant others'. (Anyone who doubts the force of group beliefs, but has no wish to trudge through the reported studies, should look no further than

Alison Lurie's most excellent of novels 'Imaginary Friends', itself a thinly disguised account of a very famous piece of research.) (13) There is no doubt that only the rarest of individuals are so deeply convinced of the truth of their claims that they can persistently maintain their views in the total absence of intersubjective legitimation. In our society such people run a considerable risk of being labelled schizophrenic and incarcerated in mental hospitals. There is that much obligation attached to intersubjectivity.

2 Congruence with preconceptions

Knowledge-claims must broadly fit what the claimants preconceptions already allow, and, in line with (1), such preconceptions are invariably part of a shared sub-cultural matrix. In Holzner's suggestive term, an 'epistemic community'. (14) Of course any such community may institutionally allow for the possibility of innovation – new knowledge-claims which are negotiable among members. But, with certain important exceptions to which I shall return, such knowledge-candidates must meet some criterion of congruence with pre-existent 'established' knowledge. The piece of knowledge may be new just as long as it does not constitute a devastating contradiction of well-entrenched views. Radical change occurs only infrequently, though important innovations are possible within at least some of these 'conservative' sub-cultural matrices.

3 Demonstrability

My first two criteria have related directly to other people and to other knowledge: an epistemic community involving its own specialised sub-culture. Demonstrability adds the view that, if our knowledge-claims are to be accepted, we must be able 'publicly' to demonstrate their truth. What will count as a demonstration may vary from sub-culture to sub-culture, as will the meaning of 'truth'. One classic scientific view (a product of a particular epistemic community) sees demonstration in terms of observable phenomena where 'observable' has a very strong objectivist and externalist slant. Obviously this is an unnecessary limitation. Provided that we see the terms <u>in conjunction</u> with (and inseparable from) the criteria invoked in (1) and (2), we can safely think of demonstrability in terms of correspondence to 'perceived' phenomena. But the claim that this criterion stands among the others is essential, for it locates the 'rules' for correspondence in a socio-cultural context, and lends no absolute status to either 'truth' or 'perception'. I should also note that demonstrability is often an 'in principle' question; an individual may never be called upon actually to demonstrate anything.

Once met, in whatever specific form, these three criteria provide members of an epistemic community with grounds for accepting knowledge-claims as true. That seems fairly unexceptionable. But so far this has been cast very much at a common-sense level, seemingly unphilosophical. How, then, does this view relate to philosophical theories of truth and knowledge? Needless to say there can be no short answer, but one or two points will bear making. The classical views of truth are commonly divided into 'correspondence' and 'coherence' theories. Correspondence theories claim something to be true if it corresponds to a 'fact'; coherence theories, more or less, that truth derives from the coherence of a knowledge-claim with other accepted knowledge. There is a third, perhaps less easily defined position, which has been added in recent years: pragmatism. Though I am not entirely sure how distinct this position is (on questions of truth at least), it broadly claims that what is true is what works, whatever gives good practical results.

All three approaches find some expression in my commonsense criteria. That seems only right; common sense has generally been more comprehensive than philosophy! Demonstrability involves all three, in that correspondence, coherence, and practicality must all be demonstrable, though there is obviously a specification of demonstrability which has particular affinities with the correspondence theory. The crucial qualification is that this is a relativised version of correspondence, one in which the 'correspondence rules' are specified by the relevant sub-cultural matrix; there are no universals involved. Thus a member of a mystical cult who claims 'truth' to be attainable through application of certain rituals does invoke a correspondence between his or her views and the 'real' world, but with a very special set of correspondence rules. My second criterion (congruence with preconceptions) is obviously a direct reflection of the coherence theory, though also in relativised form. Once more the intersubjective context defined in criterion (1) is the source for 'rules' of coherence and an institutional setting for already established knowledge. Different epistemic communities develop different specifications of the three general criteria. As to the pragmatic approach, I rather think it can be argued that this is no more than 'correspondence' with a special set of rules. Something (an account of a phenomenon) is true where it gives good working approximations in a range of situations. The correspondence rules relate theoretical fictions indirectly to a presumed 'real' world, and those rules are once more a part of the sub-cultural matrix. So, my commonsense view incorporates elements of all the classical approaches, rather than committing itself to one or another.

I do not borrow the traditional terminology of the theory of knowledge to lend this discussion spurious philosophical legitimacy, but simply to indicate some of the ways in which my common sense epistemology borrows from all approaches rather than leaning on any one. Given my generally antiabsolutist perspective this is hardly accidental. In the course of a useful discussion of 'truth' Hamlyn remarks:

> What is important about most of the expositions of the classical theories of truth is that they are part and parcel of the classical theory of knowledge, the search for certainty. (15)

It is because I am not engaged in any such futile search that my three criteria cross-cut the boundaries of correspondence and coherence. Absolute truth is a non-starter; truth is something we arrive at in certain ways and according to certain 'rules'. It is only necessary to limit those 'rules' in very specific ways (e.g. correspondence or coherence) if we are in the business of prescribing the irreducible oughts of scientific inquiry. It is unfortunate that the sociologist's penchant for tossing philosophical labels about has recently accepted a simple dichotomy in correspondence/coherence terms. I hope it is clear that, as Hamlyn notes, it is only in the search for absolutes that we require such exclusive distinctions. Elsewhere they are of little value other than as polemical weapons.

So far, then, I have suggested three types of criteria essential to the common-sense process of developing 'true knowledge'. The substance of these criteria remains open, for they may vary from context to context. In short, they provide a basic framework within which different epistemic communities develop their own particular specifications. It is arguable that western culture is quite homogenous in this respect; certainly I am inclined to think so. However, such a claim is not essential to this argument, and I can safely continue without it. The question which arises now is: what sort of processes are involved in our creating knowledge? It is the word 'create' that should be stressed because arriving at knowledge is indeed a creative process; knowledge is not something that the world passively delivers up to us. Instead it is the product of a series of creative interactions between knower, known, and the human and natural context within which both are located. Once more there is a philosophical abyss at hand. This time, however, I can only deal with the issue by making bald declarations here, and filling them out later in the book. So I shall present a simple list of the features of everyday inquiry, though it is important to remember that this involves enclosing a process of continuous interaction in an unavoidably static framework. I hope the usefulness of its simplicity outweighs the unavoidable distortions.

For convenience I divide the process into elements, though they obviously blur together at their borderlines. Loosely then:

(a) We <u>experience</u> some aspect of the world which, for whatever reason, is significant and/or problematic to us.

(b) We <u>interpret</u> this experience in the light of whatever preconceptions we feel able to apply to it.

(c) The <u>derivation</u> of these all-important preconceptions lies in our earlier experiences, our intersubjective situation, our interests, our social world, in what we have already accepted as 'knowledge', and in a variety of other personal and situational factors.

(d) We <u>test</u> these interpretations against our further experiences and against those we derive, directly or indirectly, from significant others.

(e) In this way we arrive at socially <u>established</u> knowledge (if you prefer - intersubjectively established knowledge), sometimes referred to as 'truth'.

(f) This 'truth' may then be used to <u>explain</u> our further experiences. We use it to make sense of our world and, in this way, knowledge begets knowledge in its own image.

Several qualifications are necessary. First, the idea of 'experience' is not without problems, and it is certainly not easily separated from 'interpretation'. It seems likely that the two are conflated in our normal thought processes, and that, in many ways, interpretation is constitutive of experience. But it is useful to distinguish the two if only to underline the fact that habitually we do try to separate our actual interpretations from the putative something 'out there' which we experience. Second, the idea of 'testing' these interpretations is not intended to be as rationalistic as the term itself might suggest. For one thing this is not an atomistic claim: not all statements are tested, at least not directly. In addition, since my scheme does not specify a definitive meaning of 'test', (d) serves only to suggest that we do attempt to relate our interpretations of empirical events to other such events, usually to convince ourselves and others of the cogency of our particular case. In this characteristic fashion we try to demonstrate that our views are something more than merely subjective. Within different epistemic communities (science, for example) we find different criteria by which 'adequate testing' is recognised; these 'rules' will figure prominently in later chapters. Third and last, it is crucial to recognise that establishing knowledge is not simply a question of the intellect, but always depends to some degree on social, psychological, and cultural factors. Indeed, it is this 'non-rational' element which lends much knowledge its remarkable resilience in the face of contradictory evidence. Of course, if our knowledge persistently failed - if, in Stoppard's delightful phrase, our maps proved to be a 'conspiracy of cartographers' and we were forever walking over unanticipated cliffs - then finally we should be forced to abandon it. (16) But 'finally' may be a long time coming, and history is replete with examples of entrenched knowledge surviving many a tumble over unseen cliffs.

THE SCIENTIFIC CANOPY

Allowing that this account has some plausibility, and also allowing that science is not so radically different from common sense, what image of 'scientific' inquiry can we now develop? Quite obviously the three criteria still apply, for science is a human activity like any other. Thus:

1 Attaining 'scientific knowledge' depends, in part, on intersubjective legitimation by significant others, i.e. by members of the 'scientific community', though there are occasional cases where the response of a more general public can outweigh such restricted expertise. Some such legitimation is a necessary, if not sufficient, precondition for knowledge to be credited as 'scientifically acceptable'.

2 Candidate knowledge must fit in with the general expectations of the scientist and his or her significant others. In effect, a considerable proportion of scientific activity falls within what Kuhn has termed 'normal science', bounded by a 'paradigm', and involving scientists in a process of diligent 'puzzle-solving'. The paradigm defines the puzzles, the form of available acceptable solutions, and the substantive and methodological views to be 'taken for granted'. Not that all scientific practice is 'normal science', and Kuhn recognises the existence of anomalies as well as more extensive innovations than the simple paradigm model might suggest. He conceptualises these characteristics in terms of the shift from paradigm to paradigm: scientific revolutions. Science changes when one paradigm is overcome by anomalies and another replaces it. Though very influential, the detailed adequacy of this analysis is debatable; I shall return to it short-

ly. (17) But it does serve to remind us of the largely 'conservative' nature of practised science.

3 Scientific knowledge must be made demonstrable in terms of some intersubjectively agreed methodology, involving either or both of 'testing' and 'explanation' (in this sense 'explanation' is a test of utility). Like any other subculture the scientific community thus develops 'rules' for recognising acceptable knowledge, though, less typical, science appears to place some premium on this methodology being rendered explicit. This is both a virtue and a burden.

These criteria, then, reflect the general social and common-sense character of scientific inquiry, and, as elsewhere, they emphasise the need for reflexive consideration of scientific activity. Ultimately the six process elements and these three criteria will lead me to a picture of scientific activity; to a systematic reconstruction, and hence an epistemology, of science. But, first, I must raise some difficulties, many of which recur throughout this book, and all of which find some discussion in the philosophical literature. The key problems derive from the view that science is a social activity like any other and thus subject to similar 'irrational' constraints and virtues. So far I have cast that view in terms of Thomas Kuhn's work because he is probably its best-known contemporary proponent: 'The Structure of Scientific Revolutions' has been enormously influential in a variety of disciplines. But his analysis - though clearly correctly aimed - is open to much criticism. Some difficulties lie in questions of detail. It is not at all clear what socio-cultural mechanisms are at work in the process of paradigm-shift, or how anomalies reach the critical revolutionary point. Such problems make for interesting tasks in the sociology of science. But Kuhn has also provoked a number of more far-reaching responses in the philosophy of science, the most vitriolic of which have come from the Popperian school. As Kuhn remarks of an exchange between himself and Watkins, their views are 'mutually impenetrable'. (18) Understandably the Popperians have been largely conservative in the face of what they conceive to be a direct attack on the fundamentals of their position. But Kuhn has provoked change as well as entrenchment, best typified perhaps by Lakatos's stimulating 'methodology of research programmes', an attempt to combine older views with a modified 'relativism'. (19) And there has been a 'radical' response, too, and one which, in Feyerabend's account of it, grew out of a period of close discussion with Kuhn. If Kuhn has turned out as something of a prophet, Feyerabend seems to be more of an 'enfant terrible'. His is an aspect of the argument that will bear more exploration here. He puts forward both a critique of traditional (and not-so-traditional) empiricism, and a stimulating appeal for an 'anarchistic' epistemology and philosophy of science. (20)

Feyerabend's general argument begins with the recognition that the search for methodological certainty is misplaced, and that we cannot have once-and-for-all standards of truth.

> We can only speak of what does, or does not, seem appropriate when viewed from a particular and restricted point of view, different views, temperaments, attitudes giving rise to different judgments and different methods of approach. (21)

Obviously this invokes the same relativistic spectre we find
in Kuhn. Where the two differ is that, in Feyerabend's
view, Kuhn is too conservative, failing to see the full implications of his starting point. Thus, having recognised the
basic character of science and correctly argued for the
existence of both the 'normal' and the 'revolutionary', Kuhn
then mistakenly imposes a specific temporal order upon them.
For the most part, he argues, science is completely paradigm-bound, only intermittently collapsing beneath the weight
of accumulated anomaly, and even then only in the face of a
new and growing paradigm. While Feyerabend can go along
with much of that general analysis, he crucially disputes the
rigid ordering. For him, normal and revolutionary science
co-exist; indeed, 'normalcy' cannot possibly be as rule-
bounded as it is in the Kuhnian ideal-type. This anarchistic image can be confusing, since Feyerabend's views do
change, and it is not always clear whether he is dealing in
prescriptions or descriptions. Certainly he is making a
plea for anarchistic epistemology, as his sub-title indicates,
and that is in some part a prescriptive aim. Yet sometimes
his criticisms of Kuhn's view appear to be criticisms of Kuhn
as <u>description</u>. Or is it that he is worried only that Kuhn's
view can be used conservatively, i.e. as a way of elevating
paradigm-bound, normal, 'boring' science into the only
route for scientific progress. For there is a striking
parallel between Kuhn's image of science and the general
Parsonian perspective. Because both see culture as so
powerful their views are open to - though do not necessarily
imply - conservative interpretations. It has already been
argued (based on Kuhn) that 'pre-paradigmatic' disciplines

like the social sciences should devote all their energies to developing an exhaustively rule-governed normal science. This would be anathema to Feyerabend, and a denial of the most basic creative sector in science as he conceives it.

Even siding with Feyerabend at this level, however, leaves us with problems. Does it really follow from the fact that we cannot have 'universally enforced standards of truth and rationality' that we should therefore be 'against method'? Put thus bluntly it does not, and Feyerabend himself (in his very detailed discussions as opposed to his general pronouncements) is not so much against method as against particular philosophical rationalisations of scientific methodology. His lengthy analysis of counterinduction, and his emphasis on the problem of incommensurability, make it quite clear that method is important. The villain is, rather, universal and monolithic method. There is thus some ambiguity in his central argument. At times he seems to lean toward the 'abandoning method' case as that is developed by, for example, Phillips; crudely, the argument that full recognition of Kuhnian relativism can lead only to discarding all methods. (22) At other times he is much more subtle, effectively arguing for a multiplicity of cross-cutting perspectives, the counter-positioning of which gives rise to the truly creative element in scientific inquiry. Indeed, without such multiplicity, creativity would not be possible, which is why the Kuhnian model fails to deal adequately with the process of change. It follows, of course, that for Feyerabend the choice between basic cosmologies is a matter of taste, but that does not mean that such choices are not open to argument. It is possible to choose between

'comprehensive ideologies', and to do so in the open.
Whether that is a 'rational choice', however, is another
matter. In his illuminating and entertaining running debate
with Lakatos (not to mention his war with some of the other
Popperians) Feyerabend frequently argues that choices at
this level cannot be 'rationally' made; we should not try to
dictate the terms on which scientific 'progress' takes
place. (23)

I have digressed into Feyerabend at such length because I
believe his to be a more persuasive variant of scientific
relativism than the much better-known Kuhnian version. No
doubt he over-emphasises the role played by 'anarchic' and
'playful' elements in science, but I am just as certain that
Kuhn under-emphasises them. But Feyerabend's less monolithic analysis seems more realistic, and his wish for a
'liberated' epistemology and methodology is far more refreshing than the many varieties of restrictive determinism
to be found in the philosophy of science. Unfortunately,
there are other, dare I say, traditional questions on which
he can be of little help. In fact, if we were to take him to
be 'against method' pure and simple (which I do not), such
questions would no longer be of any importance. I am
thinking, for instance, of the sorts of difficulties which surround 'demonstrability', a criterion universal to science in
general, but open to considerable variation in particular
specifications. Even given my relativistic emphasis, it is
here that the most serious difficulties are raised for sociology, and it is here that the still important classic problems
are to be found. Problems about 'inductive inference',
'observation languages', 'correspondence rules', and about

the 'theory laden' character of observations. Recognising that scientific activity is culture-bound does not allow us to avoid such issues. It is becoming far too common to find sociologists claiming that the recognition of relativism in science justifies our forgetting about method, either as practised or proposed. Feyerabend's discussion of incommensurability should serve to remind us that such views do not follow of necessity. Recognising that we cannot be dealing in absolutes transmutes the traditional problems, but it certainly does not eliminate them.

These qualifications stated it is now possible to give a diagrammatic statement of the 'structure' of scientific inquiry. Its baseline is relativistic; it is a system of inquiry bounded by a culture, an epistemic community, a paradigm, a disciplinary matrix, or whatever of these general terms seems most appropriate. It organises the common-sense elements of inquiry, as I have earlier presented them, beneath a 'scientific canopy'. (24)

This diagram is very much a 'snapshot'; it abstracts a set of elements and relations from a complex process of interaction, and can thus all too easily lead to misunderstanding. For instance, I do not mean to suggest that each individual scientist, in the course of a given investigation, performs all the actions implicit in this structure. He or she does not necessarily order the world, derive a theory from this ordering, test the theory, use it to explain further phenomena, and so contribute to the on-going accumulation of knowledge. Like everyone else, scientists specialise; the scientific culture incorporates a body of (at a particular time) given theory; particular methodological 'rules' may

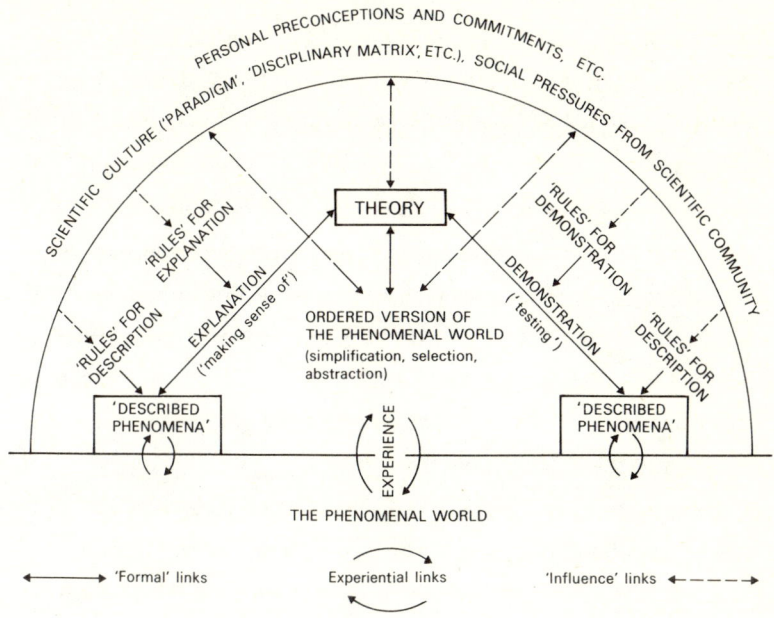

Figure 2.1 The scientific canopy

change over time; different elements of the process may receive varying emphases; and so on. These areas, and many others besides, have provoked a substantial literature, and I shall look at a number of them in detail elsewhere in this book, including, most prominently, questions of demonstrability and explanation. These are traditional areas of dispute in sociology. But there are other topics for discussion which arise equally obviously from the figure; I shall mention only two here.

The category 'theory' has been left deliberately amorphous, though it is absolutely central. Chapter 4 will discuss it in some detail. Another amorphous category which serves only to suggest a number of richly complex problems is 'described phenomena'. As employed here the term suggests that we do not test (or use) theories in relation to 'real' phenomena, dealing instead with descriptions already mediated for us by a language – a set of rules, conventions and terms that generate such descriptions. The traditional philosophy of science discussed this sort of question in relation to 'observation languages' and their relation to theoretical terms. I do not wish to prejudge that discussion in the figure: the arrows connecting 'theory' and 'described phenomena' are two-headed, and there are doubtless complex relations between these links and the 'languages' presumed to underly the descriptive process.

Obviously, then, Figure 2.1 does little more than set a scene: a scientific canopy sheltering some of the more important elements in the process of inquiry. Thus sheltered, particular sub-cultures of scientists are able to negotiate the detailed rules of the game, detail which gives specific meaning to the general processes of Figure 2.1. So, although I am suggesting that the general form of empirical inquiry is invariant (an explicitly systematised version of everyday empirical inquiry), detailed specifications can vary. There are no absolute rules, no method as such, just as there is no single scientific paradigm. There may be several scientific canopies, cross-cutting each other in one or more ways, operating at different levels of abstraction, and dealing with varied aspects of what is presumed to be the same phenomenal world. To persist with diagrams:

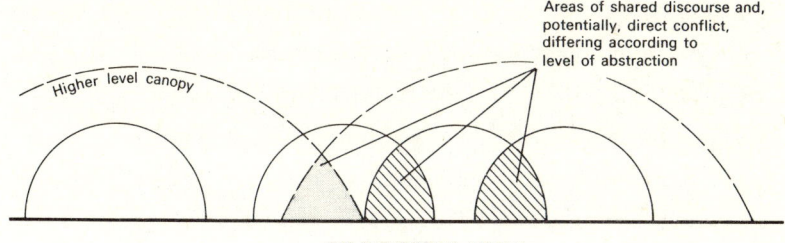

Figure 2.2 Multiple canopies

Feyerabend (whose diagrammatic conventions I have mercilessly adapted in Figure 2.2) suggests, put into the terms I have been employing here, that it is often where the provisions of one canopy encounter another that we find scientifically creative developments. For example, where the characteristic observation language from one is used counter-inductively in relation to a theory from another. (25) It is from such confrontations – at varying levels – that major scientific advances are generated, and not from the cumulation of 'normal science' or the 'all change' of the paradigm-switch.

I can take this little further now without exploring the more detailed questions. So where are we? I have argued from a simple common-sense view of empirical inquiry to a potentially complex account of 'scientific' inquiry; the degree of complexity itself depends on the specific sub-cultural configurations. It is that specification which now needs exploration. But before moving on there is something to be gained by asking and answering one more very general question: how does this position relate to traditional reconstructions of scientific method? If we reflect briefly on the three conceptual-empirical links outlined in Figure 2.1 ('experience', 'demonstration', and 'explanation') we find that various philosophies of science have elevated one or another of these elements into the defining characteristic of empirical inquiry. Thus, for example, naive versions of inductivism (and some not so naive) have emphasised the 'experience' link at the expense of the rest. They argue, in effect, that the scientist looks closely and methodically at the world, generalising from this experience to a final, acceptable, and true account. Or, as in the traditional Popperian emphasis on conjecture and refutation, we find philosophies which focus primarily on 'demonstration': isolating the problem, setting up hypotheses, and subjecting them to exhaustive test. Or, finally, there have been those rational reconstructions which place major emphasis on 'explanation' as the central goal of scientific inquiry. This has historically led to the over-deductivist view found in Hempel's early work, though much qualified later. (26) All these philosophies reify one element of the process of inquiry, and, accordingly, none of them is satisfactory. But they are certainly not to be casually discarded at the whim of sociological fashion. In even exploring only one aspect of inquiry they have focused rigorous attention on crucial problem areas. It would be a very limited discussion that now defined them out of court, and I shall surely need to lean on them later.

THE WEB OF BELIEF (27)

I have intended that this chapter provide a general framework for what follows; the problems it identifies, and the terms in which they are identified, structure the rest of this book. It has sketched a view of scientific inquiry which, like common sense, creates and elaborates on a complex web of belief. This image is broadly in line with contemporary relativised views of knowledge and science, though it does not accept – as some do – that we must therefore resign ourselves to methodological disinterest. Ascribing truth, any concept of truth, demands more responsibility than that. Let me quote Quine who also believes that 'absolute truth' is a chimera:

> Have we now so far lowered our sights as to settle for a relativistic doctrine of truth – rating the statements of each theory as true for that theory, and brooking no higher criticism? Not so. The saving consideration is that we continue to take seriously our own particular world-theory or loose total fabric of quasi-theories, whatever it may be. Unlike Descartes, we own and use our beliefs of the moment, even in the midst of philosophizing, until by what is vaguely called scientific method we change them here and there for the better. Within our own total evolving doctrine, we can judge truth as earnestly and absolutely as can be; subject to correction, but that goes without saying. (28)

Of necessity, we work beneath a canopy, scientific or even sacred. This puts us, again in Quine's phrase, 'in the same boat', and the structure of our web of belief both confines us and evolves in relation to our prompting. So instead of seeing relativism in any absolutist undermining sense – an increasingly common view – I see it as a further demand on our intellectual responsibility. Though our reflection cannot possibly escape all webs, we must be self-conscious about methodological and epistemological issues in a positive frame of mind. There is no excuse here for a retreat into methodological solipsism, a denial of all claims because it is impossible absolutely to affirm one.

Scientific knowledge is larger, then, than the elements that I have suggested make it up. That is why the 'web of belief' is such an apt description. We have become used to an image of science cast in terms of rigorous theorising, hard testing, and methodical explanatory analysis. Of course, such practices have their role; but they do not exhaust what we must understand if we are really to be responsibly reflexive about our discipline. The web which we spin far transcends the direct experiences of those who create it; the statements which constitute it do not always relate to its observational periphery. Let me quote Quine once more, though this time the question is one to which I shall return. To my mind no one has put the position more succinctly.

> The totality of our so-called knowledge or beliefs, from the most casual matters of geography and history to the profoundest laws of atomic physics or even of pure mathematics and logic, is a man-made fabric which impinges on experience only along the edges. Or, to change the figure, total science is like a field of force whose boundary conditions are experience. A conflict with experience at the periphery occasions readjustments in the interior of the field.... But the total field is so undetermined by its boundary conditions, experience, that there is much latitude of choice as to what statements to reevaluate in the light of any single contrary experience. No particular experiences are linked with any particular

statements in the interior of the field, except indirectly through considerations of equilibrium affecting the field as a whole. (29)

But it is the relation between this field, this web, and the human world which must concern us. Only given such understanding - a very small part of which is the subject of this book - can we truly claim to be responsible seekers for knowledge.

The argument that has led me here can be broadly reconstructed as follows:

1. The creation, systematisation, and development of scientific knowledge is a human process like any other.
2. As such, it shares a basic epistemological framework with 'common sense', a framework which can only include absolute terms like 'truth' and 'falsity' because such concepts are given meaning by its very existence.
3. It is possible to isolate the basic elements of common-sense empirical inquiry (intersubjectivity, congruence, demonstrability), and the elements and relations of the process of inquiry (experience, interpretation, derivation, test, establishment, explanation).
4. These elements are also to be found in science, and this scientific mode is relativistic without being retreatist, and general without being completely vacuous.
5. The questions of procedure thus raised may be answered differently within different 'scientific canopies', but the overall form of empirical inquiry is standard. Emphases vary.
6. In this way we can isolate a series of crucial problem-areas, all of which require some attention if we are to

formulate anything more than an entirely partial and misleading account of our own practices. They include, most prominently:

(a) the nature of theory
(b) the nature of experience and observation
(c) the nature of demonstration
(d) the nature of explanation
(e) the nature of the relation between scientific canopies

7 Any particular sub-culture of inquiry develops its own solutions to these problems, however insufficient these solutions may appear. Through the methodological practices thus created - though not always recognised for what they are - scientists spin their web of belief.

In the remainder of this book I shall take up these issues in one or another guise, and always with the framework of this chapter in mind. I cannot keep repeating these qualifications so this discussion will have to stand as a general qualification to later more limited discussion. In chapter 3, then, I shall consider the 'Received View'. In chapter 4 I will try to fill out the idea of 'theory'. Chapter 5 looks at the problems of explanation and chapter 6 at demonstrability. Other problem areas, 'experience', 'observation', and the collision of 'scientific canopies', recur throughout. They cannot be given separate discussion within this small compass. Finally, I shall attempt the hopeless task of drawing together the threads of my own web.

chapter 3

Theories and things

> The time has now come to move beyond the static, 'snapshot' pictures of scientific theory to which philosophers of science have confined themselves for so long, and to build up a 'moving picture' of scientific problems and procedures, in terms of which the intellectual dynamics of conceptual change in the sciences will become intelligible and the nature of its 'rationality' apparent. Stephen Toulmin, 'The Structure of Scientific Theories'

Though it does reflect my sympathies, this opening quotation is intended to underline the inadequacies of this chapter and the next. Toulmin's 1969 call to battle is admirable, even if the philosophy of science is only trundling slowly in his direction. In fact I could hardly not concur with his view since much of what I have already said implies some such shift. But, at the risk of repetition, I must emphasise that 'traditional' approaches are not to be casually dismissed. It is true that these villainous philosophies of science have been misleading to innocent sociologists, though it might also be said that natural scientists would have suffered similarly had they paid more than scant attention to their philosophical companions. But, suitably qualified, the traditional analyses can survive as part of the process of understanding scientific inquiry. They remain a useful baseline from which advance and debate can continue. That, at least, is my belief.

For the sociologist there are many ironies here. Any sociological reader of Toulmin's passage will have seen a transmuted version many times before. Sociological theory has been frequently castigated for offering structural 'snapshots' instead of 'moving pictures' of social process. But just as sociologists have argued that the snapshot is an essential precursor to the moving picture (if we are to understand change, we must know what it is that is changing), so the philosopher of science might justly claim that some necessarily static account of scientific structure is essential to process analyses of practised science. The vital misconception, in both fields, lay with those who felt that static analysis of structure exhausted the proper discourse of the subject. They assumed that the problems thrown up by their analyses – and they did identify important problems – were the only problems worth considering. This sort of reification produced the cruder _isms_ of both disciplines: in the philosophies of science, the absolutist versions of empiricism and deductivism; in sociology, the excesses of vulgar functionalism and even vulgar Marxism. Their claims were certainly important, but not exclusively so.

There are other more satisfying ironies. It must surely give any sociologist a certain malicious pleasure to observe the current state of the philosophy of science. For many years sociology trod this wilderness in search of a methodological saviour, finding several, and misinterpreting them all. Now this seemingly most elegant and agreed of subjects is more than a little dishevelled. To read the recent volumes derived from conferences and symposia is to realise how divided the camps have become; the many variations of logical empiricism which grew in that fertile Viennese soil are fragmenting. Though no one seems to be turning Kuhn back on his own discipline, there can be no doubt that the 'paradigm of paradigms' is finally on the move. Groups and sub-groups are proliferating. (2)

In the essay from which I derive my opening quotation, Toulmin distinguishes four responses to this fragmentation, ordering them by the 'amount' of logical empiricism they reject. The least revisionist group, he says, are those who still accept that the logical structure of scientific theories is axiomatic, but do not concur with the purist ideal of Unified Science. The second group includes those who accept that the essence of scientific inquiry lies in logical structure, but not necessarily of an axiomatic form. The third group weaken this still further: here scientific theories are recognised to have an 'intellectual structure', but without the traditional emphasis on mathematical and logical form. This is probably the closest to the sort of views so far embodied in my discussion, especially that derived from what I personally find most significant in the works of Kuhn and Feyerabend. Toulmin claims that Kuhn's contribution to the Structure of Scientific Theories symposium demonstrates how close he is to traditional logical empiricism, on some issues at least. Feyerabend, however, he would probably place in his fourth group (though I would not) which includes those who believe that:

it is no longer useful to discuss the concepts and hypotheses of the natural sciences in terms of 'structure' or 'system' at all, let alone in the static terms associated with terms like 'logical structure' and 'axiomatic system'. (3)

The crucial factor that differentiates the first three from this final group is that the former still allow us some degree of systematisation: they still permit 'systematic reconstruction'.

I have argued that such reconstruction must be qualified by locating it firmly in a socio-cultural context. Hence my schematic discussion in chapter 2. From this point of view science is seen as part of the world, not something superior to it, though precisely what part still remains unclear. What I have outlined as the 'scientific canopy' is a general image, necessarily low on detail. It could hardly be otherwise since much of that detail is stipulated by whatever passes as 'science' in particular sub-cultural situations. But Figure 2.1, even as it stands, does distinguish two types of relations between 'science' and the world on which it preys. One group revolve around 'informal' pressures: social, psychological, and cultural factors which partially serve to define the operations of science. The second more 'formal' set focus on topics conceived by the traditional perspectives as truly constitutive of science, and often, though not necessarily, formulable in strictly logical terms.

These two types, and argument about their relation to each other, figure prominently in the currently confused state of the philosophy of science. The total traditionalist entirely rejects 'informal' relations as significant considerations;

the total radical is similarly disposed to the 'formal'. And where the one ascends to god-like absolutism, the other sinks into a relativistic mire. Luckily most discussion falls somewhere in between, striving to find an acceptable combination. So, for example, while someone of Kuhn's persuasion seems happy to define a range of 'informal' issues as the proper domain of the sociology of science, Toulmin prefers to include all such questions within the philosopher's ambit. (4) Like everyone else, I have no simple solution. In chapter 2 I have already sketched a framework which emphasises the importance of factors alien to the logical empiricist tradition. Within the limits set by that sketch I do find 'systematic reconstruction' and a more 'formal' discussion useful; hence my present intention to explore some of the problems generated in the philosophy of science 'received view'.

THE RECEIVED VIEW: BASIC CHARACTER

It was in his much quoted paper What Theories are Not that Hilary Putnam first used the expression 'received view' in reference to the philosophy of science's then established conception of scientific theory. (5) Most other contributions to this 1960 International Congress on Logic, Methodology, and Philosophy of Science were still largely operating within the received view, though here - as elsewhere - there were signs of increasing dissent. By 1969, with its symposium on the Structure of Scientific Theories, the trickle had become a flood, and Suppe's book-length introduction to the proceedings of that conference ably charts the enormity of the change. (6) So the 1960s saw rather more than widespread 'public' recognition of Kuhn's controversial views on science; they were also the occasion for deeper soul-searching by one-time proponents of the traditional perspective. One admirable characteristic of this discussion has been its honesty and clarity: Hempel, especially, has been willing to discuss and revise continuously. (7) Of course, such perpetual willingness to backtrack may stem from a series of last-ditch attempts to save the old philosophy; it has certainly been suggested. But this would be a cruel characterisation. To someone accustomed to the imprecise rhetoric of much sociological 'debate', this period of philosophical controversy sets exemplary standards.

But that is not my reason for invoking the discussion here. I believe that it is important for us to understand the nature of our theories (and hence knowledge), not only as human products, but also as sets of ideas exhibiting some form of intellectual structure. Hence my declared interest in 'systematic reconstruction'; what might be termed, a 'moderate explication'. That aim can only be furthered by exploring some of the issues raised in recent debates about the role and status of theory, a topic widely misconceived in the sociological literature. In chapter 1, for instance, I remarked that the sociologist's concept of 'empiricism' was something of a bucket concept, and that even where it did invoke specific philosophies it neglected modern, post-Carnap, developments. Sociologists typically took to themselves - for criticism or uncritical adoption - a crude empiricism. In the rare cases where the questions were pursued further, it was a disjointed and fragmentary chase. Thus, what Hempel once called the 'theoreticians' dilemma' entered sociological discourse only as an argument between those who believed that theory did have a part to play in sociology and those who thought that the (usually survey) data spoke for itself. (8) The real problem raised by the 'dilemma' (what is the justification for theoretical terms? If they serve their purpose they are unnecessary; if they do not, then they are also unnecessary) bypassed sociological discussion. And much more important, Hempel having successfully dissolved the dilemma, consequent questions about the precise nature and function of theories were largely ignored. Only the wilder extremes, such as operationism, enjoyed brief vogues as sociologists' straw men.

This is why Putnam's 'received view', and its subsequent fate, is something that should be more familiar to sociologists. Unfortunately, it is not easy to characterise, and this discussion is bound to become rather tangled. Suppe, whose overview tries to cover the widest range, distinguishes no less than an initial and a final version, recognising that both views can and do coexist with a series of intermediate possibilities. (9) However, it is possible to begin by outlining a sort of median position, artificial though that may be. Putnam puts it thus:

> theories are to be thought of as 'partially interpreted calculi' in which only the 'observation terms' are 'directly interpreted' (the 'theoretical terms' being only 'partially interpreted' or, some people even say 'partially understood'). (10)

He then considers the way in which the received view divides the non-logical scientific vocabulary into observation terms and theoretical terms, the former referring to the domain of public observables while the latter do not. This allows for a concomitant division into observational, theoretical, and mixed statements. The whole thing fits together in this way:

> a scientific theory is conceived as an axiomatic system which may be thought of as initially uninterpreted, and which gains 'empirical meaning' as a result of specification of meaning <u>for the observation terms alone</u>. A kind of partial meaning is then thought of as drawn up to the theoretical terms, by osmosis, as it were. (11)

The usual formal shorthand for this account sees scientific theory as an ordered pair of sets of sentences. Using Hempel's nomenclature: (12)

$$T = [C, R]$$

where C represents the calculus (the axiomatised deductive system), and R the 'correspondence rules' which govern the process of partial interpretation.

A sociologist might now be forgiven for asking, 'What has this to do with me'? For one thing, sociological theories are rarely axiomatised deductive systems, even if some sociologists have intermittently demanded that they be thus formulated. (13) For another, this account was specifically intended as a reconstruction of natural science, and even then might not be recognised as such by many natural scientists! So why bother? Now I do not subscribe to axiomatisation or deductivism in sociology, at least not in exclusive roles, but that is hardly the critical point. The received view claimed to be dealing in explication: discovering the allegedly minimal elements of order to be found in scientific usage. And in this endeavour one problem rose to dominance. This was the old question of the relation between 'empirical phenomena' and the terms in which we seek to understand them. Most crudely, how do we relate our ideas to the world we believe we are analysing? And that is an important question for sociology just as it is for every other discipline. Though the received view may have got it wrong (for science and sociology), exploring exactly where it went wrong can still be constructive. In this context Suppe usefully compares the initial and final views (his 'TC' is the equivalent of Hempel's 'C,R'):

> Initially the Received View was an account of theories which attached little importance to the theoretical apparatus, TC, its function being little more than a means for introducing mathematics into science. In its final version theories are construed realistically as describing systems of non-observables which relate in incompletely specifiable ways to their observable manifestations: as such the theoretical apparatus is central to its analysis, and the emphasis is on how the theoretical apparatus connects with phenomena. (14)

Initially, then, the calculus (primarily derived from pure mathematics) was linked to the empirical world via conventionally accepted correspondence rules. From this 'pure' position, Suppe argues, the received view reacted to the challenge of the variation in actual scientific practice by including more and more qualifications and complications. Finally, it became a complex reconstruction – if, on some accounts, insufficiently complex – built in the sort of terms Putnam suggests: a calculus, a set of correspondence rules, and, in some versions, a not too clearly specified 'model'.

This construal of 'theory', which arises as an attempt to clarify the relation between that which is observed and the conceptual apparatus of the observer, undoubtedly fails to solve the problems with which it begins. What it does do, however, is point up the areas of difficulty which have to be faced. Though the terminology varies, there seem to be seven of them common to the literature (there is an eighth, 'explanation', but I shall be considering it in some detail later). One of these areas is very general, raising what might be termed 'external' issues: the status and meaning of concepts like explication and rational reconstruction. The others all focus on 'internal' issues though they can be sub-grouped. Two of them are concerned primarily with questions of theoretical form; the other four with the complex links between concept and phenomenon. In summary:

A EXTERNAL ISSUES
   1 Explication, rational reconstruction, and the general adequacy of the received view's account of science.

B INTERNAL ISSUES
   1 Theoretical Form
      (a) axiomatisation, formalisation, and deductive structure.
      (b) the nature and function of models.
   2 Epistemic Relations
      (a) partial interpretation.
      (b) problem of terms (observational/theoretical distinction).
      (c) problem of sentences (analytic/synthetic distinction).
      (d) correspondence rules.

The rest of this chapter will touch on them all, though, in accord with the general pattern of discussion in the literature, most space will be devoted to B2. Before continuing, though, I should emphasise that this discussion is intended merely to indicate broadly the ways in which the received view is considered to have failed, not so much as a self-contained critique but as essential groundwork.

THE RECEIVED VIEW: PROBLEM AREAS

To begin briefly, then, with A1. Here we encounter very general questions about the validity of the whole endeavour. What is the point of rational reconstructions such as these? Do they really add anything significant to our understanding? I have already paid some attention to such questions, and I do not propose to re-enter that discussion here. The utility of what I have been calling 'systematic reconstruction' depend on what claims are made for it. As I have observed it is inevitably partial; it selects and emphasises some factors at the expense of others. It is not an account of what scientists actually do, but rather a way of isolating the intellectual structures involved in scientific inquiry. Whatever versions of those structures it produces must be seen within their larger socio-cultural context, though that is more difficult than it might at first seem, for it involves combining the insights of two <u>apparently</u> incommensurate perspectives. Systematic reconstruction, then, is neither fish nor fowl. It combines the appearance of constructing testable theories about scientific procedure with the seemingly 'given' analytic claims of philosophy, and its products lie somewhere in the undefined hinterland between the two.

This very ambiguity makes it difficult to judge 'explicatory' claims. Suppe tries to assess the received view in terms of its adequacy as a general analysis of scientific theories, though he does interpret that question in fairly limited terms. (15) It needs to be said, however, that no systematic reconstruction can constitute an adequate general analysis of scientific theories, though it can be seen as a useful part of such an attempt. For this reason many criticisms which claim quite correctly that 'it doesn't happen this way' are not quite as damning as they may seem. In the long run, of course, I would like to see an account of inquiry which was indeed an adequate reflection of actual practice. Systematic reconstruction is part of that, but not a final answer: a partial method for self-reflexive understanding. There are critics who would abandon all such reconstructive attempts beneath the banner of total 'realism'. But a totally 'realistic' approach is just as untenable as pure philosophical analysis, for it too must abstract, simplify, and reconstruct. Seen heuristically, then, systematic reconstruction is a useful tool which can and does afford us genuine insight. We should not discard too lightly the fruits of other people's labours. But complete adequacy is not to be expected from such an approach. If it aids the process of understanding then it is adequate, if only for a while.

Turning now, then, to the 'internal' problems. This area has two domains. I shall invoke B1 (theoretical form) only briefly here, though I shall return to it later. The problems of B2 (epistemic relations), on the other hand, will occupy much of the rest of this section. (16) But first some brief 'scene-setting' comments on B1, (a) and (b). As all the commentators are agreed the 'pure' form of the received view puts major emphasis on the axiomatic deductive structure of the calculus. For the committed exponent of <u>rational</u> reconstruction the fact that few theories are thus formulated is not too important. As Hempel remarks:

> such ... observations represent no telling criticisms, I think, for the standard construal was never claimed to provide a descriptive account of the actual use and formulation of theories by scientists in the ongoing process of scientific inquiry; it was intended, rather, as a schematic explication that would clearly exhibit certain logical and epistemological characteristics of scientific theories. (17)

That is at it may be. I am not so sure that these criticisms are to be ignored that easily, though, in line with my previous comments, if axiomatisation proves heuristically useful then it has some justification. And that is exactly Hempel's problem for, as he goes on to demonstrate, formalisation and axiomatisation shed little light on the crucial problems raised by the received view. Any given theory allows a number of alternative axiomatisations, and the postulates of one in particular may not correspond to the 'basic assumptions' of the theory-as-used. In short, axiomatisation reveals nothing about the meaning of theories, and accordingly its explicatory use is limited. On deductive structure, though, there are other arguments to be considered, and I shall leave them for later and fuller consideration.

It was perhaps the very restrictiveness of the pure, axiomatised conception that led some revisionists to add a third component to the calculus and the correspondence rules: the model. In Nagel's well-known metaphor, this was the flesh on the skeleton of the abstract calculus. (18) But apart from playing this rather subsidiary role - an inferiority reflected in the fact that the usual formal accounts of the received view conceptualise only calculus and rules - models in scientific inquiry were not prominently discussed until recent years. By 1968, however, Achinstein was able to describe the subject as 'thorny'; he is given to understatement! (19) The potential and actual role of models still remains unclear, and much of the literature retains an understandable taxanomic bias. The fact that this particular topic is a late starter is to be expected, of course, in a philosophical tradition which began life by finding theoretical terms an embarrassment let alone whole models, but it is a failing of the received view for all that.

Indeed, it is this 'embarrassment' with theory which brings me to the major area of difficulty within the received view: the cluster of questions surrounding the much elaborated and much disputed theoretical/observational distinction. With hindsight one is tempted to ask why this hard and fast dichotomy was invoked at all. Putnam suggests that, especially in Carnap's seminal formulation, it was created to solve a non-problem: how to interpret theoretical terms, (20) an issue directly related to B2 (a), the doctrine of partial interpretation. Partial interpretation was intended to rescue earlier conceptions of the relation between theory and observation, but as its past exponents freely admit, it is not at all clear what the phrase means. Perhaps, because of the connotations of 'partial', it does seem to be a more realistic formulation of the relation between concepts and the 'real' world than that originally advanced by empiricist philosophy, but even that mild claim does not bear too much scrutiny. Classically, empiricism tried to reduce all terms and statements to the level of observables, where 'observable' was defined in directly experiential terms. Thus one of Quine's famous two dogmas of empiricism is <u>reductionism</u>: an ill-founded doctrine which claims 'that each meaningful statement is equivalent to some logical construct upon terms which refer to immediate experience.' (21)

The emphasis on 'meaning' is crucial. The initial problem was to discover what gave meaning to our concepts, and the various empiricist schools gave their various answers. For example, a verificationist like Carnap invokes verificational procedures as a method of linking strictly 'meaningless' concepts with the empirical world which is presumed to be the sole source of their meaning. This conflation of a term's meaning with the operations by which it is related to the empirical world raises obvious problems, and Carnap, at least, set out to solve them. In cruder versions, however, one could find doctrines like operationism, in which the meaning of a concept was entirely given by the operations used to measure it. Thus, there might be as many concepts of, say, temperature, as there are methods of measurement. Even the more sophisticated verificationist views could not escape the difficulty directly. If it is <u>meaning</u> which is given by the operations of verification, then a falsified statement is also thus meaningless. Hardly a persuasive account of scientific practice, even with the most catholic view of reconstruction!

But if I am not careful I shall be guilty of erecting straw men, and so subject to my own earlier strictures. Radically reductionist views did not survive for long in logical empiricism. Concepts, it was soon argued, were only partially interpreted in relation to observations, though, even here, as Quine suggests, the reductionist dogma survives in a more subtle form: in the claim that the isolated statements of a theory can be sensibly confirmed or denied. That is a claim to which I shall return. For the present note only that the introduction of partial interpretation gave the received view something of a let-out. A meaning thus arrived at may be enlarged, retained, or restricted as required, and to make this evasive action work modern empiricism advanced a series of 'languages' or 'vocabularies' based on a clear-cut distinction between the observational and theoretical domains. So, the received view elaborates on the basic $T = [C, R]$ by postulating a language L, a logical calculus K, observational and theoretical vocabularies (V_O and V_T), sub-languages L_O and L_T, and various more or less complex ramifications of this style of analysis. (22) In avoiding the more obvious inadequacies of radical empiricist reductionism, the received view thus erects a complex edifice in order to account for meaningful relations between 'theories' and 'facts'. The meaning of theories is now given via an extended chain stretching upward from the world of experience, via L_O and the correspondence rules to L_T, the process Putnam describes as a kind of osmosis. The 'partial' in 'partial interpretation', thus, is an imprecise defence against the absolutist claims of radical reductionism, designed to avoid commitment to an unacceptably naive theory of meaning but without losing the all-important empiricist reference. It is this attempt that necessitates the bifurcation between observational and theoretical.

This dichotomy, then, is basic to the received view; without it we cannot transcend the unacceptable restrictions of earlier empiricism. But it still embodies - if in attenuated form - one undesirable feature of the empiricist heritage. That is the apparent insistence on defining the meaning of terms by relating them, however indirectly, to a rigidly delimited empirical world. As so many critics have unflaggingly pointed out, the links between a conceptual system and the world to which it relates are far less simple than this, and, as Quine's subsequent work has amplified, the dogmas of empiricism serve to conceal the much more holistic way in which our theories are understood and affirmed. (23) But this apart, it is still possible that the observational/theoretical distinction has some heuristic value that will survive its dubious origins. Might it yet be one element of the received view worth retaining in some form or other?

The current of opinion seems to be running against this possibility. The problems are most clearly seen where the rigid distinction is made at the level of 'terms', and they are generally variations on the common-sense question, 'what counts as directly observable'? If that question cannot be sensibly answered for all cases, not just some, then the strong observational/theoretical distinction is not tenable. It does not then follow that the distinction must be discarded; it may still have some limited value. But such doubt should provide us with considerable cause for concern for, though

most distinctions can be made, not all can be made to work.
Take the paradigm examples found repeatedly in the literature: 'red', for instance, is put forward as a typical observation term, while 'electron' is seen as typically theoretical.
Plausible enough on the face of it, but only if you are prepared to accept a certain idea of what constitutes direct experience. And if that is to be so, we must also accept yet another variation on the empiricist metaphysic which the received view has already tried to avoid in developing the idea of 'partial interpretation'. The case depends on fixing a particular set of limits to 'observation' which means assuming a sensationist-empiricist metaphysic (already seen to be inadequate) or accepting the fact that such distinctions are <u>conventionally</u> set.

And so they can be. A sociologist, for example, will rightly include people's subjective beliefs within the domain of 'observables'. This, however, is not evidence for the claim that there is a radical disjunction between natural and social sciences, a claim which, in this context, would depend on the belief that natural science can operate with an unproblematic concept of 'observable'. But outside of a simple sensationist empiricism, there are no unproblematic observables for the natural sciences either; we all face a problem of complex interpretation.

But that is a digression, however important. It remains to ask whether the theoretical/observational distinction still has any heuristic utility, once given that we are dealing in questions of <u>convention</u>. For instance, one might make two illustrative lists (as does some of the literature) claiming that henceforth, for whatever purposes, the terms thus listed would be thus classified. But that would be a vacuous exercise, bearing no philosophical fruit, and having no general explicatory function in relation to scientific practice. For, as critics have frequently observed, paradigm terms like 'red' may be used in reference to unobservables without any necessary change in meaning. So even applying a more limited conventionalised division within specific epistemic communities would not guarantee that we could deal in mutually exclusive categories. Terms do not have to be either theoretical or observational. The line is blurred, and applying a clear distinction seems to raise more problems than it solves. And if that is the case at the level of elementary terms, the difficulties must be compounded with sentences and, indeed, languages. So, the clear distinction between L_O and L_T - essential to the doctrine of partial interpretation - is far from safely established.

Clearly B2 (a) and (b) are interdependent. The received view requires the doctrine of partial interpretation if it is to give observational meaning to the terms of its calculus; partial interpretation requires a clear theoretical/observational distinction if it is to meet the demands the received view is making of it. Perhaps there are still senses of 'partial interpretation' which can be saved? The majority of terms invoked in empirical inquiry may be said to be only 'partially interpreted' in that they are often open to further expansion of definition and application. But this relatively unimportant and literal version is a long way from the received view's empiricist 'grounding' of terms in an observable world. And if the difficulties raised so far are thought insufficient, then the third problem area, B2 (c), adds yet more. This time, however, the arguments are a little more devious, operating both in general philosophical and particular philosophy-of-science terms. I shall consider them in relation to the two best-known contributions: Quine's general discussion of analyticity, and Putnam's related critique of the analytic/synthetic distinction as applied to science. (24)

We have already seen something of Quine's argument about empiricist reductionism. He also speaks of analyticity as a 'dogma' of empiricism, finally reducing the two dogmas to the one view. To begin to outline this complex argument - which is all I can possibly hope to do - I must first try to develop a simple account of the received view's use of the analytic/synthetic distinction. The schematic outline I shall offer is a long way from Carnap's sophisticated considerations, but it will serve to suggest the nature of the case. (25) So, heavily pruned, the basic dogma might be expressed thus:

1 If a statement is to be accepted as scientific ('cognitively significant') it must be possible to declare that statement to be true or false.
2 A statement may only be said to be true or false if, and only if, its truth/falsity can be shown to be consequent upon either its logical structure combined with the meaning of its non-logical terms or its adequacy with respect to direct observation.
3 In philosophical shorthand, a statement which may be said to be true or false must be either analytic or synthetic; it cannot be both or neither.
4 Thus, for a statement to be accepted as a scientific

statement it is a necessary though not sufficient condition that it be either analytic or synthetic.

It is not too difficult to see how such a dogma could become part of the empiricist received view. Always concerned to provide some empirical guarantee for the elements of scientific theory, exponents of the received view had to face the fact that not all statements accepted by scientists as science were amenable to direct observational verification. Wishing, however, to limit scientific statements to those that could be declared true or false, they imported the analytic/synthetic distinction. Synthetic statements are no problem: they are open to direct observational verification. But as long as all the rest are analytic then, once having accepted the doctrine of partial interpretation, the meaning of these statements is also observationally controlled. So, as long as a statement fell into either class it could be safely construed as scientific, and Quine is clearly quite right to see the dogma of analyticity to be closely intertwined with that of reductionism.

Before going on to outline Quine and Putnam's detailed arguments, there are two very obvious general comments to be made. The first is simply to note the atomised conception of scientific knowledge which finds expression here. The dogma of the analytic deals in statements (sentences) and statements only; it assumes that science is exhausted by such formulations, and it tries to ensure that each individual statement is open to arbitration as to truth. Both the arguments I am about to consider end up seeing scientific knowledge rather differently from this. The second comment concerns the concept of truth implied here. As one might expect from empiricist reductionism, 'truth' is defined in terms of a somewhat restricted correspondence theory. As I have already tried to suggest, this is hardly an adequate account (by itself) of the truth criteria at work in both everyday and scientific practice. It is a product of misplaced demands for certainty. So, in terms both of 'atomisation' and of 'truth', there are already storm clouds gathering on the analytic/synthetic horizon.

The storm breaks with the contributions of Quine and Putnam. The outline account I shall advance here, however, does not stand as an argument in itself; for that, the reader can do no better than look to the originals. But for those unable to do so, this cryptic resume may prove helpful. Roughly, Quine advances the following case. (26) We can unproblematically claim that there are analytic statements which are logically true, whatever our interpretation of their particular terms. This is the case because of the conjunction of their logical operators. His simple example is: 'No unmarried man is married' - a true statement regardless of the construal we put upon 'man' and 'married'. However, there is a second class of allegedly analytic statements which does raise problems because the exact nature of their analyticity is not clearly specified: the exemplification here is 'No bachelor is married.' This is not a logical truth as it stands, but it can be made into one by replacing 'bachelor' with its synonym 'unmarried man'. Analyticity, therefore, is no longer the problem - it is synonomy that is problematic. And, as Quine contends, synonomy is no better understood than the initial concept of analyticity. The grounds on which we are justifiably to translate 'bachelor' into 'unmarried man' are all such as to render the statement less than analytic. Obviously they depend on taking definition and translational stances inimical to the absoluteness implicit in concepts of the analytic.

Finding no satisfactory solution to the problem of synonomy consistent with the tenets of empiricism (a problem he pursues elsewhere in his work as well as in Two Dogmas of Empiricism), (27) Quine returns to analyticity. This time he takes up Carnap's solution (embedded in the received view) which is cast in terms of semantical rules and artificial languages. In this version a statement may be said to be analytic for a certain language, a language at least partly defined by an associated set of semantical rules. But again this is no solution as such, for we are no longer talking of analyticity per se but of something different: analyticity-for-a-language. In effect the question is removed one step further, this time into the realm of semantical rules. And they are no better understood than was analyticity in the first place. In sum, then, Quine is arguing that it is impossible to specify a non-problematic concept of analyticity to be used as a fundamental element in our reconstructions of scientific inquiry. The use of the analytic/synthetic distinction in empiricism is indeed a dogma.

Inevitably Quine's paper provoked much debate which I shall not be considering here. It seems a fair summary to say that his case has been widely accepted if only as a demonstration that no one has yet provided a workable empiricist conception of the analytic/synthetic distinction. Quine would claim that the distinction is untenable anyway, but it is at least arguable that this case has not been satisfactorily made. But even that more qualified conclusion rocks some of the foundations of the received view. Quine, incidentally, takes the matter further in the search for a non-dogmatic empiricism. He suggests that, since the truth of statements does obviously depend on their language and on 'extra-linguistic fact', there arises a natural, if unjustified, tendency to believe that all statements must be analysable into factual and linguistic components. But this is a mistaken belief (also to be found in the received view) and has given rise to much confusion. Accordingly he claims,

> science has its double dependence on language and experience - but this duality is not significantly traceable into the statements of science taken one by one. (28)

and goes on to present an image of knowledge which only provides single testable statements at its 'boundaries'.

This more holistic conception of empirical knowledge is also emphasised in Putnam's discussion. (29) Unlike Quine, he is not so much concerned with the general philosophical legitimacy of the analytic/synthetic distinction as with demonstrating that most statements in practised science cannot sensibly be construed as either analytic or synthetic. An analytic principle he takes to be a statement that could not be rendered false unless the meaning of one or more of its terms was altered; clearly a restatement of the received view's emphasis on analyticity as logic-plus-meaning. Synthetic statements, as always, are those that may be directly verified or falsified. These definitions given, he then seeks to demonstrate that many vital scientific statements are cast in terms of 'law-cluster concepts'. Such concepts are not given meaning through the received view's 'partial interpretation', but by virtue of the place they occupy in a cluster of scientific laws. In short, it is the laws collectively, not individually, which give such concepts their meaning. Hence the name 'law-cluster concepts'. Forgetting any unease we might feel about the concept 'law' (for the term itself is not essential to the argument), a sociological example of such a law-cluster concept might be 'class'. Its meaning derives from a whole range of sociological statements in which it is embedded, not from a single direct relation to the empirical world.

Putnam continues the argument by pointing out that any individual general statements embodying 'law-cluster concepts' may be denied without the meaning of the concept changing; since it is not defined by partial interpretation, but by the rest of the cluster, its meaning remains relatively stable in the face of falsification. This is obviously a realistic view which Putnam buttresses with a number of examples. So, principles embodying law-cluster concepts are not analytic, and Putnam's general image of science is heading in a similar direction to Quine's. But perhaps these principles are synthetic and all can yet be saved. Once more through his detailed examples Putnam argues that many such principles are not refuted by particular individual experiments and observations. So they are not synthetic either, and we are left with serious doubts about the analytic/synthetic distinction and hence the whole received view.

By now it should be clear that there are important difficulties with the received view in B2 (a), (b), and (c), rooted, arguably, in Quine's two dogmas. For <u>reductionism</u> is no more than the claim that all the extra-logical <u>terms</u> of a theory must be linked to direct observation, and <u>analyticity</u> requires that all <u>statements</u> of a theory must be directly or indirectly connected to the empirical world. It is in this sense that the two dogmas are virtually one. And these views are sufficient to require, in their turn, problematic inventions like 'partial interpretation', observational and theoretical languages, and, indeed, 'correspondence rules' to hold the whole rickety structure together. In fact, B2 (d) (correspondence rules) can scarcely be seen as a separate problem area; it is either an extension of the other three (suffering from all the same difficulties), or a general label for all the B2 problems. It is no accident that the 1966 Minnesota Conference on Correspondence Rules should actually produce a series of papers on the problems of the whole received view. (30) And it does seem a little optimistic of Hempel to claim that a significant proportion of the received view may be rescued by admitting that 'correspondence rules' are 'theory-laden', and renaming them 'bridge principles'. (31) The various criticisms (including some I have not considered and some I have yet to mention) are surely too strong to permit a relatively simple rescue operation. In effect, the received view assumes a radical distinction between matters of fact and matters accepted as conventionally true. I argued in chapter 2 that such clear distinctions were epistemologically inadequate, imposing frontiers where there is only no-man's-land. Accordingly, I can end this part of my discussion with a quotation from Quine who sums it up far better than I:

> The lore of our fathers is a fabric of sentences. In our hands it develops and changes, through more or less arbi-

> trary and deliberate revisions and additions of our own, more or less directly occasioned by the continuing stimulation of our sense organs. It is a pale grey lore, black with fact and white with convention. But I have found no substantial reasons for concluding that there are any quite black threads in it, or any white ones. (32)

We need a 'systematic reconstruction' to fit that finely evocative description, one that avoids the more obvious traps of the received view and can help us make more intellectual sense of the sociological enterprise. (33)

The as yet incomplete argument of this chapter may be broadly reconstructed as follows:

1 Though it is vital that we understand the 'process' of scientific inquiry, some understanding of structure is first essential.

2 The major area in relation to which the structure of scientific inquiry has been explored is 'theory'. The main philosophy of science discussion has been cast in logical empiricist terms giving rise to what Putnam has called the 'received view'.

3 As a point of departure (and to understand some of the limited versions imported into sociology) the contemporary debate over the status of the 'received view' is extremely useful.

4 The received view conceives scientific theory in terms of an axiomatic calculus and a set of correspondence rules which relate the calculus to the 'real world'. In some versions an additional concept of 'model' is included.

5 There appear to be seven basic problem areas in the received view:

A.1   Explication and rational reconstruction
B.1 (a)   axiomatisation and deductive structure
B.1 (b)   models
B.2 (a)   partial interpretation
B.2 (b)   terms (observational/theoretical distinction)
B.2 (c)   sentences (analytic/synthetic distinction)
B.2 (d)   correspondence rules

6 Detailed consideration of these problem areas in the literature (and, derivatively, in this discussion) leads to the conclusion that the received view is a much too limited and hence inadequate reconstruction (even if systematic reconstruction itself is legitimate).

The problem is now to construct a useful notion of 'theory', which can be sensibly applied to sociology, and which avoids as many of these pitfalls as is possible. This is the subject of chapter 4.

chapter 4

Elements of theory

> Those who see a model as a mere crutch are like those who consider metaphor a mere decoration or ornament.
> Max Black, 'Models and Metaphors'

Most of the convolutions found in successive versions of the received view were occasioned by the need to understand 'theory'. Any analysis of science which is strongly rooted in empiricism is bound to have difficulties here, difficulties even in answering the primary question: why do we have theories at all? As the received view changed over the years its proponents, having accepted the necessity for theoretical terms in science, found it increasingly difficult to fit that recognition into their general reconstruction. The apparatus of observation languages and partial interpretation, though designed to give us some grasp of the ways in which theory functioned, served only to over-simplify and distort a complex situation. The root of the problem lay with the central empiricist act of faith: commitment to the view that science's achievements stemmed from its constant reference to the empirical. If theoretical terms were necessary, then it was also necessary to ensure that they were safely anchored in a world of observables. Concepts might skate about like so many sailing dinghies on the surface of an ocean, but not without a long cable mooring them to the solid bed beneath.

In retrospect it's easy to see the limiting effects of that commitment. The history of the received view can be conceived as a long struggle to overcome the influence of its basic dogmas, a struggle which resulted in one conceptual oddity after another. I have explored some features of that history in chapter 3, enough, at least, to suggest a perspective progressively collapsing under the weight of its own amendments. But it was not only in its compromises and extra complications that the received view proved wanting. Led into difficulties by its basic empiricist commitment, it was led into yet more by trying to preserve that commitment at each and every turn. This meant that some topics were set to one side, some 'solutions' to received view problems largely ignored. One such concerned the status and function of models in scientific inquiry. How did models relate to the other elements of theory conventionally given pride of place in the received view? Were they constitutive features or merely helpful analogies? What part did they have to play in giving meaning to the theoretical terms that proponents of the received view had found so troublesome? These and other questions were for many years relegated to virtual insignificance while the received view developed in more acceptably empiricist directions. It was only with increasing interest in the 'metaphorical' approach to science (arguably owing most to Max Black and to the analysis built on Black's foundations by Mary Hesse) that the true importance of the 'model' idea became apparent. (1) It cannot be said that any final analysis has been achieved; the literature is bewildering in its variation. But it has surely established that no account of scientific theory can even begin to grapple with the central problems without giving prominence to models.

That said, I should add that this chapter will not focus centrally on models. That topic will recur throughout the rest of this book, but here I have more schematic and less ambitious goals. In the outline structure of Figure 2.1 the focal category, 'theory', was left unexplicated. That was deliberate, but now - having considered some of the failings of the received view's attempt to grapple with 'theory' - I am in a position to fill in the blanks. Exploring the received view suggests certain guidelines as to what topics must be considered and what traps must be avoided. Within that perspective I shall try to schematise the elements of theory.

LANGUAGES, MODELS AND SENTENCES

What problem areas emerge from the received view's chequered history? Here I shall single out six:

1 It is essential to recognise the inevitable partiality of all forms of reconstruction. Even where we set to one side questions of socio-cultural context, reification is a real possibility; philosophy of science history makes that all too clear. Note, however, that the most thorough-going critics of the received view are those who incorporate the socio-cultural into their analysis. Suppe groups them together as Weltanschaungen analysts, starting with Toulmin and working through to Feyerabend. (2) I sympathise with them (what sociologist would not?) but not to the

extent of dismissing the whole received view into a relativistic limbo. There are features worth preserving.
2. Any reconstruction, whatever its particular emphases, must deal with:
   (i) axiomatisation and formalisation
   (ii) deductive structure
   Their importance may have been overstated in the past, but that does not mean that they are of no significance. I shall argue in favour of a limited deductive structure, particularly in the course of chapter 5's discussion of explanation. As to axiomatisation and formalisation, it seems to me that the case against their exclusive importance is well established. In saying that I do not mean to say that I am somehow against axiomatisation and formalisation in sociology or elsewhere, though I would certainly oppose their elevation into exhaustive accounts of theory construction. Their strategic use might prove clarificatory in certain circumstances, but they are not idealised goals for all aspiring sciences. The received view's axiomatic calculus is not the central feature of 'theory'.
3. Any adequate reconstruction of science must begin to specify the part played by 'models'. As I have suggested, it is the relative unimportance of models in the received-view's account that so fatally weakens it. Thus, for example, the 'problem' of the meaning of theoretical terms, the discussion of which has given rise to so much confusion, is much less of a 'problem' if we once recognise that concepts derive meaning from a variety of contexts and that models play a vital part in that process. Or, to borrow an example from Hesse, the difficulties surround-

ing the received-view's 'correspondence rules' do not arise in the 'metaphoric view' because correspondence rules are no longer necessary in an account of science which sees only one language constantly extended by resort to metaphor. (3) This is not to suggest that the model concept is simply a technique for evading received-view problems; it has positive qualities of its own. It is, for example, one way of tapping the more holistic emphasis we encountered in the work of Quine and Putnam, and it is a route toward a more realistic conception of the functions of theory: model concepts are essential in identifying the 'mechanisms' through which the phenomenal world is understood actually to operate. (4)

4 The issues raised in relation to models also suggest a more general question: how are all and any terms made meaningful? It is 'all and any' that should be stressed to avoid suggesting that so-called 'theoretical' terms are specially problematic. There are no easy answers to this question. It was by thinking that the problem was open to simple resolution in terms of the theoretical/observational distinction that the received view found itself in such dire straits; I would not want to suffer the same fate. There is, after all, a whole philosophy and sociology of language at issue here, as well as the more specific materials of developmental epistemology and psycholinguistics. All philosophy of science presumes some such general analysis of meaning, though that can hardly form a topic for this book.

5 Any discussion of science must begin to specify what is involved in empirically demonstrating the 'truth', 'falsity',

or general utility of a theory. In doing that, however, there are traps to be avoided, not least the traditional over-emphasis on discrete tests of individual statements and the associated presupposition that we can clearly distinguish between analytic and synthetic statements. Such difficulties as these are further compounded by the fact that scientists see theories as demonstrable in terms of their explanatory utility as well as in terms of 'empirical fit'. The issues are complex and they form the dominant subject matter of chapters 5 and 6.

6 Lastly, and travelling full circle, any scheme must allow for a 'relativised' conception of the 'rules' postulated as governing relations between theory and phenomenal world. In the received view correspondence rules provided that focus, relating axiomatic calculus to observations. As we have seen many critics suggest that this analysis is deficient. Correspondence rules are 'theory-dependent' in various ways, and that fact alone is enough to make them something less than the absolute arbiters of partial interpretation. I shall return to the question of 'rules' towards the end of this chapter.

Such considerations as these, then, inform the schematic account of theory which follows. It incorporates four basic elements: language, conceptual scheme, model, and sentence system. There are several ways in which their interrelationship could be represented, but here, for simplicity's sake and for ease of exposition, I shall use an amended version of Feigl's famous 'layer-cake' imagery. (5) In fact, this cake might be better represented as a cone: viewed from above, figuratively speaking, we would 'see' a series of concentric circles, only the outermost directly touching 'reality'.

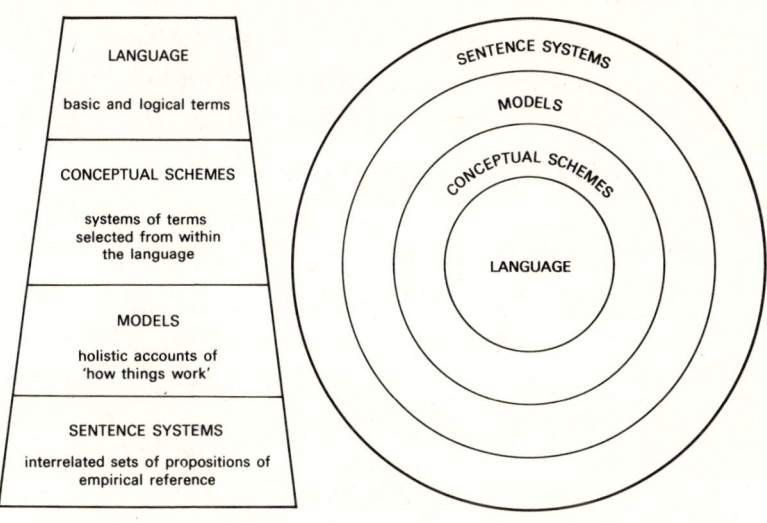

Figure 4.1 The elements of theory

Most encompassing of all, quite obviously, is the language of a theory. In the received view a great deal of effort was expended in an attempt to specify more and more closely the various sub-systems of scientific language: K, L_T, L_O and other such formulations. Since much of that work was occasioned by the felt need to distinguish absolutely between observational and theoretical modes, a similar breakdown seems neither necessary nor desirable here. Other than observing that language includes both logical operators and meaningful terms there seems no need to proliferate sub-categories ad nauseam. Why, then, distinguish a specific theory-language at all? The short answer is that there are such languages. They share terms with everyday language, certainly, but they also use some terms in special ways and add new ones where appropriate. Nor is this the case only with natural science. It could be argued, for example, that in sociology there is a more or less distinct Marxist language which makes available certain key terms like class, surplus-value, alienation, or consciousness. In that context those terms take on meanings specific to the Marxist language. Obviously such a language is not independent of the larger cultural and linguistic context; many of its terms are also available to us in everyday life. Specialist languages bear a sometimes parasitic and a sometimes contributory relation to 'ordinary' language, and there are surely complex interaction effects yet to be understood. But epistemic communities, and sub-cultures within them, do use familiar terms in distinctive ways and do generate their own vocabulary. While I don't want to carry the linguistic analogy too far, analytically isolating language as an element of theory is intended to underline that fact.

The next layer of Figure 4.1, conceptual schemes, obviously resembles language; indeed, for some purposes, they need not be distinguished. I differentiate them here, however, because I want to draw attention to the process whereby the basic terms and operators provided by the language are subject to systematic selection and emphasis. Thus, in the example of a distinctive Marxist language, we can observe a potential for several different conceptual schemes. Such systems share the language but, as in, for instance, the well-established distinction between the 'humanistic' Marxism of the early writings and the more 'mechanistic' emphases of the later work, they use it in rather different ways. Or, alternatively, let us presume there exists a general 'action theory' language. It would certainly permit a number of alternative conceptual schemes employing the basic terms (many derived from Weber) in different ways. I can envisage, for instance, at least two Parsonian sub-versions, as well as those developed by Schutz, Dahrendorf, Rex, and so many others. (6)

Conceptual schemes, then, are basic arrangements of, and selections from, the generic theory-language; taking the metaphor none too literally, they are colloquial variations. And it is within their terms that specific models are constructed: accounts of the actual workings of some phenomenon. Not necessarily in the sense that the model purports to picture real entities and their relationships, though that is one type of model. No user of the famous billiard ball account of the behaviour of gas molecules is going to suggest that molecules are billiard balls - but that didn't make the model any less constitutive of our understanding of gas behaviour. Models may be, and often are, 'imaginary' in the sense that ontological claims as to their 'reality' are not necessary for their constructive use. They are also always abstract accounts, though, depending on the language and conceptual scheme to which they relate, they may appear to be more or less 'distant' from the phenomena they purport to model. The characteristic language of Parsonian theory, for instance, appears far less directly related to the 'real world' than that found in, say, Goffman's work. To some considerable degree, however, this is an illusion, a function of the relation between language, conceptual scheme and model, on the one hand, and 'everyday' language on the other. The nearer a particular theory is to everyday usage the more likely it is that its models will be considered to be less abstract. But that would be misleading. While we might argue that a model is more or less useful, or that it 'fits' reality reasonably well, whatever its appearance it is always an abstraction.

In the growing literature on models in science the idea of 'metaphor' has played a prominent part. (7) That is not surprising. Even in the limited model concept that emerged in the received view, it was recognised that metaphorical constructions were to be found in science. At best, however, these only involved the 'substitution' and 'comparison' views, conceptions which enabled their exponents to deny metaphor a truly constitutive role. Against them Black advances the 'interaction' view (derived from Richards) wherein models may be seen as 'speculative instruments', and where new meanings emerge from such metaphorical speculation. (8) In this account, metaphor, and the model in which the metaphorical intention is expressed, is an irreducible feature of scientific inquiry; not just the 'mere crutch' of my opening quotation.

I shall not discuss the operation of metaphor in any detail, nor shall I seek to summarise the variously elaborate classifications of models now to be found in the literature. For the present it is sufficient only to note that a model is a sort of simulacrum of the phenomenon it models: simplified, selected, often fictional (at least initially) and sometimes borrowed from other domains. Any language, unless it is very limited indeed, will permit the formulation of a wide range of models putatively modelling both the same and different phenomena. Equally, alternatives can be found within single conceptual schemes competing with and/or complementing one another. Note, though, that relations between models and the two 'higher' layers are not deductive in form. A model is created within a language and conceptual scheme, but it is not deduced from them. It is an imaginative construction expressed within a particular framework. Note, also, that models come as bundles, not individual statements. Their proponents may loosely claim that they 'fit' the facts, that they are widely accepted (intersubjectively plausible), even that they are consistent with certain evaluative assumptions, but they are rarely tested in any direct sense, nor do they necessarily deductively generate hypotheses for test.

It may be, of course, that there are occasions when a particular model is directly assessed, or, at least, as directly as can be any conceptual product. Then, with Harré, we could sensibly talk of a shift from model to mechanism, where instead of simply postulating some machinery to make sense out of the 'facts' we feel able to make ontological claims for the model in question. (9) This may have been the case, for example, with atomic models and the seemingly endless 'discovery' of sub-atomic particles; it is a subject on which I am hardly qualified to comment. There also may be some sorts of models (Freud's account of personality structure in id, ego, and superego terms) which seem quite beyond the reach of an 'ontological experiment'. In short, though models legitimately may be claimed to 'fit' the facts in toto, their relation to 'reality' is rather devious. There is no reason even to suppose that we can straightforwardly distinguish between 'fictional' and 'real' <u>models</u>, or that it is necessary always to do so. The speculative element of inquiry given form in the construction and use of <u>models</u> is not, and could not be, rigorously delimited. Perhaps that is why it posed such difficulties for the over-formalistically inclined received view.

None of which is to suggest that there is no process of 'testing' in the traditional meaning of that term: hypothesis, prediction (or retrodiction), and test. Any reconstruction that failed to allow for this possibility clearly would be lacking. But the hypothesis-test can be, and has been, over-emphasised, for it is but one element in the general process of demonstration. The plausibility of theories depends on all sorts of factors, not only demonstrated consistency with the accepted <u>sentence systems</u> of a science. But it is here that we can locate that boundary of which Quine writes, the shifting surface at which a theory 'touches' reality. <u>Sentence systems</u> are interrelated sets of statements, some of which may be discretely intelligible statements as to what is the case (low-level empirical generalisations), while others, because of the terms in which they are expressed, might not be amenable to such conceptual isolation. Those involving Putnam's 'law-cluster concepts', for example, could not be empirical generalisations in this simple sense. If we are to believe the Willers, (10) much sociology deals in 'systematic empirical generalisations', that is, with low-level and restricted <u>sentence systems</u>. And there is some force to their claim, although the importance to sociology of such 'law-cluster concepts' as 'class' does surely suggest that we have progressed beyond naive systematic empiricism. As to the form in which the statements of a <u>sentence system</u> are interrelated, it may indeed be deductive, but to suggest that it is so exclusively (as did some traditional views) is positively misleading. With all layers of the theory-cake it is 'consistency' which best describes the pattern of interrelation, both between and within the elements.

Given this style of reconstructing 'theory' it is possible to imagine a specific formulation in which <u>models</u> are restricted to axiomatised calculi, <u>sentence systems</u> exhausted by sets of empirical generalisations stated in an observation language, and 'consistency' defined in terms of so-called 'correspondence rules'. In other words, the received view is one restricted possibility conceivable within this framework; needless to say, there could be others. And this, surely, is what one asks of a systematic reconstruction - that it allows for wide variation in actual practice while isolating the systematic features that underly such variation. Even with these four elements, potential patterns of interrelation are complex. Within the limits of diagrammatic conventions we can see something of that in Figure 4.2:

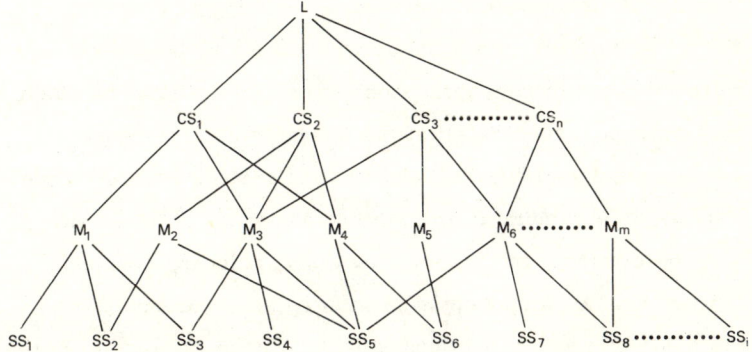

Figure 4.2 Interrelations among theoretical elements

THE RULES OF INQUIRY

Compared to the received view, what is missing from this account of theory is the correspondence rules. Long the subject of debate and disagreement, they initially featured as the rules which governed the manner in which the axiomatic calculus related to the empirical world. Later, of course, that relation was reformulated as 'partial interpretation' and, later still, Hempel renamed the correspondence rules 'bridge principles' in an ill-fated effort to include the idea of their theory-dependence. But whatever the variations within the received view - and there were plenty - there was always some attempt to formulate rules of correspondence. In this discussion (particularly in chapter 2) I, too, have made reference to 'rules', though hardly the correspondence rules of tradition. In Figure 2.1, for instance, there is provision for 'rules' governing modes of 'demonstration' and 'explanation' and for the generation of what is to count as 'data' or 'described phenomena' for a particular epistemic community. In the received view explaining, demonstrating, or observing are, so to speak, the givens of analysis; correspondence rules are needed only to 'solve' the problem of the empirical meaning of concepts. From the point of view I am developing here, however, the meaning given to concepts is not something to be understood as a rule-governed relationship to empirical referents, either directly or via the 'osmosis' of partial interpretation. Meaning is a product of the interrelationship between the various elements of theory, inevitably including some empirical reference: a concept's meaning, that is, derives from the system in which it is embedded. Hence the lack of correspondence rules in this discussion.

But I have nonetheless used the term 'rules', and at this
point it is appropriate to make some general observations
about that usage. Firstly, then, the term itself is misleading. Indeed, I have only used it because it is so well
established in the literature. In suggesting that there are
'rules' relating to various features of inquiry I am trying to
reflect the fact that inquiry is ordered in some way. In
other words I am not suggesting that there are clear, explicitly formulated rules to which scientists always attend
in the process of their work. That would be a foolish suggestion since it is so evidently not the case. As Feyerabend
has observed:

> The idea that science can, and should, be run according
> to fixed and universal rules, is both unrealistic and pernicious. It is <u>unrealistic,</u> for it takes too simple a view
> of the talents of man and of the circumstances which encourage, or cause, their development. And it is <u>pernicious,</u> for the attempt to enforce the rules is bound to
> increase our professional qualifications at the expense of
> our humanity. In addition, the idea is <u>detrimental to
> science</u>, for it neglects the complex physical and historical conditions which influence scientific change. (11)

We can, however, express a sense of the internal order of
empirical inquiry by talking as if it were rule-governed.
The task of analysis is not to formulate a set of once-and-for-all methodological regulations; specific practices will,
after all, vary from epistemic community to epistemic community. In using 'rules' I am wanting only to suggest that the
practices in question (explanation, demonstration, data
creation, and the like) are ordered, and that the precise

character of that order will vary from case to case. So there is no question of coming up with a set of methodological absolutes and, even where practitioners profess to follow certain explicit rules of inquiry, it may be that their actual practice is ordered differently.

Reading 'rules' in this limited way, then, what can we say in general about their function and about the type of rules to which we need to pay attention? If there are no independent and absolute rules of scientific inquiry, on what does practice depend? Here it may be worth distinguishing between two types of 'theory-dependence'. It seems reasonable to suppose that implicit in a theory's language are, at the very least, certain restrictions on the sorts of models that might be developed and the sorts of materials that are permitted to feature in the statements of a sentence system. In choosing a particular theory, that is, we also choose to be methodologically restricted in ordered and discernible ways. The rules which express this order are thus theory-dependent in a sense internal to 'theory' as it appears in Figure 2.1. But there is also a sense of the dependence of rules where the reference is to features of the epistemic community and its 'scientific canopy' rather than to the inherent restrictions of a particular theory-language. The areas to which I have already drawn attention seem to me to have rules which are dependent in this fashion. Offering explanations and demonstrations is ordered in ways profoundly influenced by the workings of the community in question, ways which transcend specific theories. Our theories are 'suspended' beneath the scientific canopy in a web of rules and practices which will vary between and within disciplines. Later I shall have cause to discuss sociology in terms of this variation. And though it may not always be possible to distinguish 'internal' and 'external' dependence in practice (this whole schema, remember, deals in analytical abstractions) the distinction is worth making. It points toward areas of interaction between so-called formal and informal elements of inquiry which we need to explore if our understanding is not to be unbalanced.

The historic contribution of philosophy of science in this area, however, has been unbalanced. Sometimes it has sought to define absolute rules of scientific inquiry, and even where it has had the more limited goal of rational reconstruction there has been some tendency conveniently to forget the limitations intrinsic to that form of analysis. Nevertheless, because they paid attention to scientific activity as rational and systematic, the best of the classic contributions are well worth exploring. For this reason I shall look at explanation and demonstration (and by implication the sorts of areas in which our rules must operate) by exploring the contributions of two major figures in philosophy of science: Carl Hempel and Karl Popper. This discussion forms the subject matter of the next two chapters. Meanwhile, let me summarise the argument of this one:

1 The analysis in chapter 3 of the received view suggests six rough guidelines. They are:
   (a) Always recognise the partiality of systematic reconstruction.
   (b) Ensure that the roles of axiomatisation, formalisation, and deductive structure are kept in perspective.

(c) Give the concept 'model' the prominence it merits.

(d) Recognise that the problem of 'meaning' of scientific terms is not open to simple solution.

(e) Find ways of construing 'empirical demonstration' such that we neither over-emphasise the discrete 'test' nor presuppose the analytic/synthetic distinction.

(f) Ensure that the 'rules' of science are seen in their relativistic context.

2 This leads to a systematic reconstruction of 'theory' involving four elements: language, conceptual scheme, model, sentence system.

3 Theory exists within the scientific canopy in a network of 'rules', where 'rules' is a term suggesting a pattern or order but not necessarily a set of explicit formulae. Rules may vary from theory to theory and from epistemic community to epistemic community.

chapter 5

On explanation

> The most pervasive and seriously misleading myth of contemporary philosophy is the belief that the ideal form for knowledge, and particularly for scientific knowledge, is the deductive system.   Rom Harré, 'The Principles of Scientific Thinking'

Harré is surely correct in identifying 'deductivism' as the most pervasive image of the structure of scientific knowledge. Variations upon that theme are to be found throughout the received view, whether in the orthodoxy of the formal calculus or in the mixed revisions that have succeeded it. That, as Harré also suggests, deductivism is the most misleading of such myths is not quite so clear. At least two objections can be raised to such a view, neither of them in defence of deductivism as such, but both intended to recover some part of deductivist analysis from the critical dustbin. One is simply repetition of a point that I have already made: deductive structure (or something very like it) has a limited but important role in the process of inquiry. Though we must divest ourselves of the excesses of deductiv<u>ism</u>, it would be wrong to discard every aspect of that perspective. For the philosophy routinely condemned as 'deductivist', and this is the second point, is usually misleadingly characterised in such a way as to suggest that deductive structure is exhaustive of all 'proper' scientific knowledge. But there are examples of modern explanatory deductivism, influenced especially by Carl Hempel, in which the deductive ideal is less extreme. In accepting Dray's description of his perspective as 'covering law analysis' Hempel himself is endorsing a view of his own work which departs from pure deductivism, if only by recognising that deductive subsumption is merely one possible strategy for 'covering' an event to be explained. (1)

It is important to emphasise that these objections to antideductivism do not imply support for the deductivist perspective. I state them here because I believe that it is worth considering the deductivist analysis sympathetically, not least because it has been historically important in both philosophy of science and sociology. In practice that means taking yet another look at Hempel's well-known discussion of explanation - not the only version of the argument for deductive subsumption, but surely the most widely debated. The work arising from it, especially in relation to historical explanation, has been of obvious significance for sociologists. (2) Yet there is little explicit discussion of Hempel's views in a sociological context, a surprising absence in a discipline much given to such methodological debate. Apart from his analysis of functionalism (which has a different, though related, axe to grind) Hempel's influence has been tacit. (3) While many sociologists have implicitly accepted a loose deductivist ideal as distinguishing the 'traditional' conception of sociological theory, few have openly maintained such a view. Instead, where sociologists have raised the issue they have been inclined to resort to somewhat implausible vulgarisations of Hempel's position.

A clear case of such borrowing is to be found in the work of George Homans. He has tried to erect a methodological credo on the basis of a simple deductivist perspective. In two books ('Social Behaviour' and 'The Nature of Social Science') he has advanced what is perhaps the crassest account of sociological explanation currently available. (4) 'Social Behaviour' may be the more excusable. Its analysis of explanation leans not on Hempel but on Braithwaite's less subtle specification of deductivist orthodoxy, and, although one might question the validity of using a 'rational reconstruction' as a blueprint, the real arguments with that book lie elsewhere. (5) But 'The Nature of Social Science' is quite indefensible. It bases its case in Hempel and makes open reference to his more recent writings, work which, as we shall see, carries him some way from his initially pure deductivism. But all this bypasses Homans. He argues that the purpose of scientific theory is to provide explanations, and that:

> the explanation of a finding, whether a generalization of a proposition about a single event, is the process of showing that the finding follows as a logical conclusion, as a deduction, from one or more general propositions under specified given conditions. (6)

For Homans it is this crude summary that represents what is to be learned from established philosophy of science.

Yet he would be hard pressed to defend that view as rep-

resenting Hempel's position, except perhaps in relation to
the 1948 paper, Studies in the Logic of Explanation. Since
Homans himself makes reference to the later work (1965)
reported in Aspects of Scientific Explanation one can only
assume that he is aware that there are significant changes
in Hempel's analysis. In particular, that Hempel accepts
the claim that probabilistic 'laws' are of considerable importance and, hence, that 'showing that the finding follows
as a logical conclusion, as a deduction' is not a defining
characteristic of all scientific explanation. Like much
sociological borrowing from the philosophy of science,
Homans's derivations from Hempel are fundamentally misleading. Not that it would matter were he alone in advancing such interpretations. Unfortunately he is not, and the
image of explanation imputed by sociologists to philosophers
of science has all too often been naively deductivist. (7)
For this reason, and because of the more general insights
afforded by studying his work, I shall base this chapter on
a consideration of Hempel's views. Needless to say there
are other important traditions in the discussion of explanation. The fact that they do not feature here is no reflection on any contribution they might make to sociological
self-understanding. (8)

EXPLANATION AND COVERING LAWS

Hempel's initial statement on explanation was published in
1948, spelling out a view implicit elsewhere in his work
(the much debated discussion of the role of general laws in
history preceded this paper by six years), in Popper, and
in other versions of the received view. (9) In that paper

Hempel and Oppenheim concentrated attention on scientific explanation, claiming to have identified its basic logical structure. Dubbed 'deductive-nomological explanation' (henceforth D-N explanation) their analysis was designed to reveal the logical form in which scientists set about answering explanation-seeking why-questions: Why does this gas expand? Why did the Bolsheviks come to power? Why does this child fail in the educational system? In asking such questions the object of explanation is taken as given, though it should be noted that this is hardly unproblematic. After all, we do not explain events as such but rather events-under-a-description, and there may be a number of available descriptions of what is putatively the same event. We may thus offer different explanations according to the description in use. (10) But that problem can wait. For the present let me outline the basic elements of D-N explanation.

Taking E as the explanandum-sentence (i.e. the description to be explained), explanation takes the form of deductive subsumption beneath a set of laws in conjunction with a set of initial conditions. In this way E is shown to be a deductive implication of the conjunction of laws and conditions. This can be seen formally, employing Hempel's own nomenclature:

(1)
$$\frac{C_1, C_2 \ldots\ldots\ldots\ldots\ldots C_k \quad \text{Conditions}}{L_1, L_2 \ldots\ldots\ldots\ldots\ldots L_r \quad \text{Laws}}$$

$$E$$

The nature of this deductive relation between explanans

(that which does the explaining) and explanandum can be seen more clearly when reduced to its minimal elements. Assume that what is to be explained is the fact that object 'a' exhibits G-like characteristics: we wish to explain the incidence of G in the case of 'a' (Ga). The minimal D-N form would be:

$(x)(Fx \supset Gx)$ Conditional sentence (law)

(2) Fa Condition
      —————
       Ga

Or, in words, for all x's, F-like characteristics imply G-like characteristics. In the case of 'a' (a member of the class of x's) we find characteristic F. The conjunction of this empirical condition with the conditional sentence logically allows us to conclude that G will be present. And that, of course, is what we set out to explain in the first place. Or, to employ a deliberately simplified example which will recur throughout this discussion, assume that we wish to explain why a particular child has a low level of educational achievement in the English school system. A D-N explanation of this fact would utilise a conjunction of the law (true, let us say) that all working-class children exhibit low-educational achievement and the empirical condition that this particular child is working class. The particularity to be explained has been subsumed beneath the general truth as represented by a conditional sentence. Do not be misled, however, by the simplicity of this syllogistic construal. Explanations - as (1) suggests - will usually involve several laws and several conditions, those laws often stating far more complex relations than straightforward implication.

There are some aspects of this formulation which have caused unnecessary confusion. I shall raise them here only briefly; they are not central to my concerns. First, D-N analysis makes no commitment to any idea of causation between the elements represented in the law. Hempel does argue that explanation of an event by invoking its causes is implicitly deductive-nomological (in that it appeals to a general law relating cause and effect) but what is presented in D-N explanation is a logical argument not an ascription of causal efficacy. The statements of that argument may be related to such causal ascriptions, but that is neither inevitable nor even necessary.

Second, any statement of a law is itself available to be explained or 'covered' by a yet more general law or set of laws, leading to a deductive network of such law-statements. Clearly the regress of explanations is potentially infinite though, in practice, it is likely to stop at what have been referred to as 'theoretical principles'.

Third, Hempel is not claiming that all scientific explanations explicitly utilise deductive-nomological form. This could hardly be the case unless one was willing to dismiss vast ranges of scientific practice as unscientific. What Hempel and his followers have tried to demonstrate is that this logic underlies accepted scientific explanation in a wide variety of disciplines. Needless to say this claim has given rise to much debate.

Finally, and at a rather different level, there are difficulties about identifying precisely what constitutes a general law for purposes of D-N explanation. Laws, Hempel suggests, can be neatly defined as true law-like sentences, but how do we characterise true law-like sentences without referring back to the concept of law? An obvious possibility is to define any true conditional sentence as a law, but to do so would require us to consider some patently trivial explanations as scientifically adequate. Perhaps the most popular example of one such is the 'law' that 'all men are mortal' used to explain Socrates's mortality: Socrates is a man therefore Socrates is mortal. Such an explanation, although formally correct, would rightly be considered uninteresting. Yet reluctant though he would be to accept this formulation, Hempel does none the less weaken his position by permitting such dubious progeny as 'elliptically formulated D-N explanation', 'partial explanation', and even 'explanation sketches'. If the boundaries can be blurred in this respect why should they not be in relation to the characterisation of a law? Hempel himself makes only general suggestions about the precise nature of laws without ever providing a really systematic account. (11)

But these areas of confusion have not been the major loci of critical attack on Hempel and the D-N model. Such criticisms have fallen into two main groups. To the first belongs the various claims that there are legitimate scientific explanations which do not involve subsumption or 'covering'. Such arguments allege that law-like statements are not actually invoked in the course of explanation, or, stronger, that they are not possible, or, even stronger, that they are anyway not necessary. The most common context in which these claims appear is in the debate on historical explanation, often advanced by those who see history, sociology, and the other 'cultural sciences' as ideographic disciplines not concerned to achieve scientific generality. There is no need to enter into that particular debate here. The ground has been well covered in the literature and, I believe, Hempel has defended himself rather successfully from many of these charges. (12) The second group of criticisms, however, is much more immediately relevant to this discussion. These critics make the more specific claim that the logic of deductive necessity does not apply to a wide range of explanations which must none the less be accepted as legitimate. Though they are prepared to retain the idea of subsumption, it is retained without the deductive emphases of the D-N model. Scriven identifies what is, in effect, the key claim of this position:

> it is virtually impossible ... to find a single example of something that is normally called a law in science which can be precisely formulated in non-probability terms. (13)

If this is so then it must also be virtually impossible to find any examples of true D-N explanation since one can hardly found an argument for deductive necessity on a probabilistic law-statement.

Although Hempel would not accept Scriven's extreme claim (he believes that there are significant areas of science in which D-N explanation is possible) he does recognise the force of the probabilistic argument. 'Aspects of Scientific Explanation' makes it clear that he sees D-N explanation as only one form of explanation, not a complete account. Accordingly he suggests two further types of 'statistical explanation'. The first and more genuinely statistical, so-called 'deductive-statistical', need concern us little here. It is not directed toward the explanation of particular events but toward provision of a deductive argument 'explaining' a statistically formulated law. But the second type, 'inductive-statistical', is intended as an explanation of particular events. Its goal is the same as that of D-N explanation, but its laws are framed in probabilistic terms. Employing Hempel's notation we can represent I-S explanation thus:

(3) $$\begin{array}{ll} C_1 C_2 \ldots\ldots\ldots\ldots\ldots\ldots C_k & \text{Conditions} \\ L_1, L_2 \ldots\ldots\ldots\ldots\ldots\ldots L_r & \text{Laws} \\ \overline{}\; [\text{likely}] & \\ E & \end{array}$$

It is readily apparent that the 'covering' relation remains an essential part of the process, though there is no longer any deductive necessity linking <u>explanans</u> and <u>explanandum</u>. Instead, the conclusion E is considered to be more or less 'likely'. Once more that can be clearly seen in a minimal characterisation. Assume that the incidence of G in case 'i' is to be explained in I-S terms:

$$p(G,F) = r \qquad \text{probabilistic sentence (law)}$$

(4) $$\begin{array}{ll} F_i & \text{condition} \\ \overline{}\; [r] & \\ G_i & \end{array}$$

That is, there is a certain probability (r) of G-like characteristics being associated with F-like characteristics. In case 'i' we find F-like characteristics and so, with a probability of r, we have 'explained' the occurrence of G.

It will be clear that the introduction of I-S explanation is something more than a minor addition to Hempel's theory. Indeed, as I shall try to show, it carried him well away from the 'positivist deductivist' views for which he has been so frequently castigated, a fact which makes it all the more regrettable that so much attention is still paid to D-N explanation as the most 'genuine' form. Even if Scriven's claim that almost all scientific laws are probabilistic is an exaggeration (and much recent work might suggest that it is not) no social scientist could deny the importance of probabilistic statements. And lest the label 'statistical' should mislead us into thinking that I-S explanation relates only to precisely formulated statistical associations, I should emphasise that the form of explanation outlined in (3) and (4) applies more generally. Though Hempel does say that probabilistic claims attribute a characteristic to a 'specified proportion' of a class it does not follow that such specification must be given precise numerical value. He also suggests that phrases like 'almost certain' or 'very likely' commonly occur in probabilistic argument.

The key element of Hempel's analysis that is retained in the I-S version is, of course, 'covering'. For 'covering' does not necessarily require an argument cast in terms of deductive necessity, but rather that the elements of an explanation always include a combination of what Bunge has termed 'Generalizations' and 'Circumstances'. (14) To explain something is to subsume it beneath a generalisation which can be shown to apply to the particular case to be explained. The differences between D-N and I-S explanations, therefore, are not differences in relation to this general structure, but differences in the status of the generalisations themselves and in the sorts of arguments which can be based on them. D-N argument deals in <u>nomic</u> necessity, unambiguous and unarguable. Explanations involving probabilistic laws have a built-in potential for ambiguity. As Scriven has vividly suggested: 'an event can rattle around inside a network of statistical laws.' (15)

As an unrealistically simplified example take once more the question of a child's educational achievement. Assume that child 'i' has proved unsuccessful in negotiating the hurdles of the English educational system. A common response to a why-question in this case would be to offer an argument about the high probability of educational failure (F) among children of working-class parents (W) (for the sake of completeness let me put an arbitrary figure of 0.8 on this probability). Thus, 'i's' failure would be understood as follows:

$$p(F,W) = 0.8$$

(5)　W_i
　　　——————　$[0.8]$
　　　F_i

An 'explanation' which has been widely accepted by sociologists, and - as far as it goes - is still of some importance. Consider now an alternative argument whose premises we will assume to be true but which leads to the conclusion that 'i' will not fail ($\bar{F}$). Suppose that children who have no siblings (S) have a probability of 0.7 of success. Then:

$$p(\bar{F},S) = 0.7$$

(6)　S_i
　　　——————　$[0.7]$
　　　$\bar{F}_i$

There is no reason why both the 'laws' expressed above (working-class children generally do badly in the educational system; only children generally do well in the educational system) cannot coexist as part of the same body of 'true' sociological knowledge. Indeed, they have done so in the past and they probably do so now. And yet, applied to 'i', they could be used to explain both success and failure. Or, put another way, where probabilistic laws are involved, a given and not inconsistent body of knowledge can lead to inconsistent conclusions.

This sort of ambiguity (Hempel speaks of it as epistemic ambiguity) does not arise in the case of D-N arguments. It is instructive to see why. Basically it is because the relation between premises and conclusions in a D-N argument is one of logical necessity. So, a body of deductively related laws (definitionally logically consistent) cannot possibly lead to logically inconsistent conclusions. In the context of the example, the law $(x)(Wx \supset Fx)$ (for all children, working classness implies educational failure) would, if accepted as true in a particular body of knowledge, preclude the truth of $(x)(Sx \supset \overline{F}x)$ (for all children, lack of siblings implies educational success) unless, of course, there are no members of the class of x's who are both W and S. For the difficulty is one of reference classes — there are children who are both only children and working class. D-N arguments deal in conditional sentences; sentences which apply to all members of a class. The probabilistic sentences of I-S argument do not have that sort of certainty. We are not able to speak of 'for all children', but instead must qualify the class to which the putative laws apply. To avoid the sort of ambiguity illustrated in this example we would, in Hempel's terms,

> want an acceptable explanation to be based on a statistical probability statement pertaining to the narrowest reference class of which, according to our total information, the particular occurrence under consideration is a member. (16)

And that requirement makes the initial description of the event to be explained especially vital, for it is at that stage that the reference-class parameters will be set. Argument (5) might apply to child 'i' in as much as 'i' is working class, male, lives in the north of England, and has two or more siblings. But however carefully we specify the appropriate reference classes there does remain some looseness in the I-S form. Events can indeed 'rattle around'. (17)

Hempel resorts to Carnap's 'requirement of total evidence' as a way of dealing with ambiguity in I-S explanation. However, this is surely not a satisfactory solution to the problem. Unlike much of Hempel's own discussion, which applies to the formal structure of explanation, that requirement is professedly a piece of practical advice (maximise your evidence, narrow your reference class) and so guarantees nothing. There still remains the possibility of yet further probabilistic laws finding inclusion in a body of knowledge and acting as premises in arguments which lead to contradictory conclusions. Any sociologist familiar with that discipline's propensity for formulating post hoc explanations will know how easy it is to find accepted sociological generalisations to 'explain' an event whichever way it happened. It is sometimes claimed that this is simply an unfortunate feature of post hoc explanation. In fact it stems from the inherent ambiguity of what are inevitably rather loosely formulated probabilistic generalisations. (18)

I do not emphasise epistemic ambiguity in a wholly negative spirit. That would be foolish, for we have little choice but to come to terms with difficulties such as these. If more care is required in qualifying the scope of application of our generalisations by strict specification of the reference classes to which they apply, then that is the price that must be paid. Some might argue, of course, that if a reference class is sufficiently narrowed it would prove possible to express regularities in conditional form; where there are not many x's 'for all x's' become less forbidding. But, as Hempel himself observes, the statement $p(G,F) = 1$ is not logically equivalent to the universal conditional $(x)(Fx \supset Gx)$. (19) If we were to limit our attentions to such universal 'laws', we would be bound to sacrifice a good deal in terms of scope and generality. To do that in fear of ambiguity would be to cut off one's nose to spite one's face. It is more appropriate to argue that if sociology is to be other than ideographic, and if explanation does indeed involve 'covering', then we are better employed facing up to the ambiguities and problems rather than retreating into smaller and smaller reference classes. But that sort of awareness is impossible if discussion of sociological explanation in philosophy-of-science terms is limited to attacks on, or defences of, the D-N version. Sociological generalisations are unavoidably probabilistic in form. It is the implications of that which must concern us, not some limited ideal of deductive theorising.

Two issues, then, arise from such considerations. The traditional appeal of the deductive-nomological form is that it requires and creates a parsimonious, logically concise, deductively interrelated framework of propositions. This sentence system expresses the most complex of theories in a form which is both general and precise. If, for whatever reasons, we cannot aspire to such a form, if we must utilise probabilistic argument, then our explanatory accounts are inevitably incomplete and ambiguous. We have already seen something of the ambiguity; the incompleteness is more important still. It relates to the fact that a probabilistic sentence, simple or complex, is little more than a statement of correlation. To claim $p(G,F) = r$ is to claim that the relation between two variables approximates to a particular form in a particular degree. That is, the 'laws' which we utilise in explanations are, in effect, empirical generalisations.

Like their stronger counterparts in D-N argument they too require explanation. But where it can be argued that there is no need to step outside of the deductive structure to find further explanations of D-N laws (though I would not accept such an argument), no such argument can be sensibly formulated in relation to statistical laws. These probabilistic claims require 'qualitative explanation', whether that is in terms of ascribing causation, identifying appropriate mechanisms, or whatever. In short, they need to be interpreted within the ambit of a model. Though we might 'explain' a particular event by covering it with an appropriate statistical regularity, and although that regularity might be covered in turn, there remains the task of making sense ouf of these generalisations. And that understanding must come from outside the assembly of generalisations through the process I shall refer to, somewhat portentously, as 'explanation-by-model'. I say 'portentously' because the process itself is simple enough; however, it would be difficult to overestimate its importance. The Hempel account of explanation, because it was first formulated in D-N terms, underplays the importance of the mechanisms that serve to relate the elements described in the law-statements. For him, 'all references to analogies or analogical models can be dispensed with in the systematic statement of scientific explanations.' (20) The inclusion in his scheme of I-S explanation, however, does make some such consideration inevitable. Accordingly I shall now look more closely at the explanatory functions of models.

EXPLANATION BY MODEL

Consider, then, the simple example I have been using. What, normally, would we expect as an explanation of 'i' failing in the English school system? The observation that 'i' was a child from a working-class background, and that there is a high probability of such children failing, would, in itself, surely be considered insufficient. Partly, of course, because of the difficulties encountered in all probabilistic explanations of this sort - problems relating to reference classes and lack of specificity - but also because such an explanation simply doesn't go 'deep' enough. Explaining 'i's' failure in terms of an isolated proposition associating class and educational achievement tells us nothing about the whys and wherefores of the process. It enlightens us only in the sense that we now recognise that 'i's' failure falls into a class of such events: a significant proportion of working-class children fail. And that, in turn, sets us a further explanatory problem.

In the sociological literature on this topic the attempt to achieve greater explanatory depth has developed in several directions. The most obvious strategy is material in its emphasis. (21) Working-class children fail, the argument runs, because of the material restrictions of working-class life: lack of housing adequate to provide homework facilities, the economic necessity of early employment, lack of materials other than those provided by the education system itself, and so on. Such arguments are implicitly inductive-statistical. In them the original statistical regularity, $p(F,W) = r$, is broken down into a series of propositions relating features of working-class life to certain material states of affairs and, in turn, relating these material states of affairs to failure in the educational system; an elaboration of the 'intervening variables' linking F and W. In effect, the association between F and W is taken as the explanandum and is 'covered' by a series of sentences expressing probabilistic relations between factors like class background, housing conditions, out-of-school work patterns, and failure in examinations. Though nothing would be added to our understanding of working-class children's experience of education by formulating such accounts in explicitly I-S terms, they clearly could be thus formulated without losing whatever cogency they possess. And that, of course, would be sufficient to sustain a view of explanation which emphasised 'covering' as its formally distinctive feature.

But would we accept these more detailed propositions as a basis for a greater explanatory understanding of the initial problem, namely 'i's' failure? I think we might, but not because the <u>form</u> of these further explanations fits a preconceived pattern. Explanations in terms of lack of facilities and so forth are (more or less) persuasive because of the <u>substance</u> to which they appeal. We intuitively accept that homework requires quiet and privacy and that such work is necessary if a child is to be 'successful' in school education. But the plausibility of such claims is not intrinsic to them. Rather, we accept them because we tacitly accept the model of the learning process in which they are embedded. Though we may take for granted an association between homework and educational achievement, we do not do so simply because we have observed it to be the case in the past. Lurking behind such assertions is the 'common-sense' claim that more work means more achievement, and, though it is stated here as a simple proposition, that claim is inextricably bound up with a whole conception of the nature of learning. (22)

Explanatory cogency, then, requires that we 'cover' the sentence to be explained and map it into an acceptable model of the processes presumed to be at work. Only where that model is already widely accepted and 'naturalised' can we delude ourselves into thinking that explanation is exhausted by propositional covering. But the moment our 'laws' invoke something unfamiliar or counter to prevailing beliefs, then the importance of models as constituents of explanation becomes more apparent. Though explanation of educational failure in terms of poor home facilities may appear to be a simple application of common sense, there are also more culturally elaborate explanations which presuppose alternative less 'obvious' conceptual models.

Take, for example, the sort of analysis found in Bernstein's work. (23) Faced with the problem of explaining low working-class achievement in school education Bernstein follows a line of argument that would not be considered self-evident in the same way as would be the more materialist case. The locus of his account is language, and, in particular, ideas about the coding of language derived from structural linguistics. Beginning with a distinction between 'restricted' and 'elaborated' codes, Bernstein develops a model interrelating the division of labour, class, authority relations within the family, and socialisation experiences, and traces their effect on language use and relative success rates in education. One's judgment of this explanatory analysis depends as much on one's attitude to Bernstein's model of authority roles in socialisation (and its relation to English class structure) as on the efficacy of his work in providing suitable covering propositions. It may indeed be possible to demonstrate that working-class children's language-use is crucially characterised by its restricted code, and hence that they are ill-fitted for the English education system. But what is crucial to the final plausibility of the thesis is Bernstein's ability to identify the mechanisms through which linguistic codes relate to education and through which socialisation experience relates to language.

To put it more generally, a model offers us explanation in the sense that it advances particular and appropriate accounts of determinancy. A sentence system provides us with a series of propositions of varying generality expressing the relations existing between a set of elements. To that 'knowledge' the model adds an account of the mechanisms underlying those relationships; it makes them intelligible in the sense of showing why they are determinately related. Traditionally, of course, the favoured mechanism involved establishing <u>causal</u> links. It is this emphasis that allowed some exponents of D-N explanation to avoid reference to any framework beyond the 'laws' themselves. Their <u>nomic</u> character could be presumed to correspond to the constant conjunction and temporal ordering of conventional attributions of cause, and thus it could be assumed (though not by Hempel) that D-N explanation had causal attribution built into it. Hence there was no apparent need to refer to a further model in order to explain the relations expressed in the 'law'. The 'law' itself, with its implicit causality, did all the explaining necessary. But even if we were dealing in explanation by <u>nomic</u> necessity (and that is rarely, if at all, the case) <u>cause</u> is not the only form of determinancy available to us. There are a variety of ways of conceptualising the mechanisms underlying any given pattern of determinate relations. (24) And in so far as probabilistic statements are normally those at issue, even if our target mechanism was causal in character we would still need to map the relationship to be understood into an appropriate model. To say that something is causally related to something else is to frame that relationship within a perspective that permits such attribution. (25)

    Fortunately it is not necessary to pursue the question of determinancy and cause in this context. Sufficient only to establish that, although we may accept I-S explanations of particular events, that acceptance hinges on our also accepting the model in which the explanation is embedded. Clearly the kind of explanation offered by models is different from that involved in 'covering' described phenomena. While 'covering' explanation involves somehow demonstrating that the explanandum is a specific case of already established law-like statements, model explanation purports to provide an account of the mechanisms at work in the relationships that are given propositional expression. Accordingly the relation between a model and the sentences that it 'explains' cannot sensibly be described as deductive. It is a relation of 'plausibility', 'fit', or 'coherence', rather than a precise link amenable to formal expression.

Indeed, for any given sentence system there may be several candidate models. A judgment among them will no doubt depend on a number of factors including intersubjective plausibility of the model, its apparent adequacy in relation to other sentence systems deemed relevant, and its capacity to stand up to 'ontological test'. (26) However, ontological testing - the attempt to establish the 'realism' of the mechanisms proposed by a model - may well be the last move in the explanatory game, not the first. Models often begin life as fictions; explanation by model is an exercise of the creative imagination. Whether they are based on analogy, on homology of structure, or on some other of the many possibilities explored in the literature, they are inventions that we use to make sense out of something that we think we know. (27) And in that sense they are just as central to explanation as are the 'law-like' propositions of covering law analysis.

EXPLAINING AND INTERPRETING

Clearly this emphasis on the importance of models in explanation is related to the set of views developed in chapter 4, and to the growing interest in models in modern philosophy of science. Ideas about metaphor are central here. As Mary Hesse observed in the course of developing her 'metaphorical view' of science: 'the deductive model of scientific explanation should be modified and supplemented by a view of theoretical explanation as metaphoric redescription of the domain of the explanandum.' (28) It is perhaps an accident of history that it is the model that supplements the deductive system and not vice versa. The logic of the metaphoric account surely gives primacy to the model in which the metaphoric function is realised; the process of 'covering' is only an element in that metaphoric redescription.

To explain something, then, is to map one or more given propositions into what Hesse calls the secondary system, a process of redescription intended to deepen our understanding of the workings of a given phenomenon and hence 'deepen' our explanations. On this reading (and I am far from doing justice to the full complexity of Hesse's account) a judgment about an attempt at explanation is a judgment about the degree to which we can 'describe' an event in such a way that it features as a coherent element within our theory. That involves 'covering', certainly, and part of the machinery of that process is what is central in Hempel's account. But taken on its own Hempel's analysis suffers from the common failing of the logical empiricist tradition: its focus is on isolated propositions. In contrast to that emphasis many modern contributions to the philosophy of science (including authors as diverse as Quine, Kuhn, Hesse, Putnam and Feyerabend) develop more holistic theories, recognising that epistemic functions like 'explaining' or 'demonstrating' cannot be fully understood in atomistic terms. The importance of this holistic re-emphasis is more apparent in relation to our understanding of empirical demonstration, though that is a contingent rather than a logical connection. I shall explore it in more detail in chapter 6.

Meanwhile, I ought to make some brief observations about the relation of this discussion to other conceptualisations of explanation, particularly those which see the social sciences as uniquely problematic. There is a well-developed tradition which takes <u>social</u> explanation as its topic and which explores that topic in the light of diverse sources in the philosophies of action and language. Throughout this book I have deliberately eschewed any such discussions, partly because they have not figured prominently in the philosophy of science as it is conventionally understood and partly because their consideration is, for me, part of a separate though related project. I intend to continue in that vein, pausing here only to observe that I do not believe that there is any necessary inconsistency between the sort of views I am developing here and at least some of those in which the focus is on meaning, action, and interpretation. One of the virtues of emphasising the role played by models in explanation, and the holistic relation between explanans and explanandum that this implies, is to undercut the traditional emphasis on 'laws'. It has often been the apparent determinism of that term which has led to claims that social action is not amenable to 'scientific' explanation since, by its very nature, it is not governed by universal laws. Leaving to one side the rights and wrongs of that claim, a more model-centred view seems to me to avoid the difficulty while offering positive opportunities for understanding sociological practice. To explain an <u>action</u> (macroscopic regularities present different problems) is to 'map', 're-describe' or 'translate' it into the terms supplied by a particular model. It is to place it within a network of concepts and generalisations. The real arguments then are less about the form of this 'covering' process and more about what sorts of theories, and what sorts of models within theories, are most efficacious in providing explanatory bite. They are arguments about substance. Or, put another way, they are about how we should interpret social action (by imputing motives, by considering action as rule-governed, by postulating normative commitments, or whatever) and how far we can carry those interpretations in the search for explanation. To interpret an action, to redescribe it metaphorically, is to begin to explain it, and models, whether declared or not, are absolutely integral to that process. The key question, then, is not whether a social <u>science</u> is possible, but what sort of models of social action should be employed in the course of offering explanations. Our task is to discriminate between alternative theories of action rather than between different general concepts of explanation.

Clearly this view has implications for the way I would seek to understand, for example, the debate occasioned by

Peter Winch's 'The Idea of a Social Science' and associated arguments about the role of rationality in our understanding of action. (29) I do not mean that I have somehow resolved such problems: far from it. But a framework such as this, however provisional, does provide a perspective within which to place the bewildering variety of claims and counter-claims which are candidate models for our understanding of social action. As I have said the pursuit of this task is something for another occasion; nothing I have suggested here could stand as anything other than a signpost pointing the approximate direction in which I would wish to travel. The journey itself must wait. Meanwhile, it is time to summarise the outline argument of this chapter.

1 Deductivism has suffered a 'bad press' in sociology and elsewhere, a fate only partly deserved though hardly unexpected. As applied to explanation by Hempel, however, it is a more subtle analysis than is sometimes claimed.

2 Having begun by identifying D-N explanation as the basic logical form of all scientific explanation, Hempel later yielded to the criticism that there are acceptable scientific explanations which do not rest on the logic of deductive necessity.

3 To incorporate this recognition he introduces a second form of explanation: I-S explanation. Both forms rest on the idea of 'covering': the presumption that explanation always involves subsuming the explanandum beneath an appropriate generalisation. Where they differ is in the sort of generalisation to which they refer, and in the sort of argument that may be based on it.

4 D-N explanation involves a relation of deductive implica-

tion between explanans and explanandum; it deals in <u>nomic</u> necessity. I-S explanation involves no such commitment; its internal structure is probabilistic.
5. For this reason I-S explanation inevitably involves 'epistemic ambiguity'. In Scriven's phrase, events can 'rattle around' in a network of probabilistic laws.
6. Furthermore, explanation requires additional interpretation in terms of a <u>model</u>. Explanations are thus more holistic than they are portrayed to be in the deductivist account.
7. Such explanatory holism provides a point of entry (from a philosophy of science perspective) into debates on social explanation.

chapter 6

On demonstration

> our statements about the external world face the tribunal of sense experience not individually but only as a corporate body.   W.V.O. Quine, 'Two Dogmas of Empiricism'

It is in considering questions of demonstration that we normally encounter empiricism at its most pernicious. So much so that some sociologists, recently inclined toward blanket condemnation of empiricism and all its progeny, are given to attacking any interest in demonstrability as so much empiricist hogwash. That is unfortunate, but it is at least understandable. Traditional recipes for 'testing' sociological hypotheses were often empiricist in the worst sense, demanding that we root each and every 'scientific' concept in a domain of simple observables. Nor is that necessarily confined to the past. There is still support for such views in the social sciences, though perhaps more in economics and psychology than in sociology. Indeed, in sociology nowadays it sometimes seems as if the opposite rhetoric is in the ascendant, sustaining a theorising process almost wholly unfettered by empirical restraint.

Yet demonstrability is crucial. If sociology is concerned with an empirical world, however broad the definition of empirical, then sociologists must surely develop criteria through which their accounts may be judged and compared. Naturally that demand cannot be reduced to the simple idea of 'testing'. Explanatory fruitfulness, internal coherence, ontological depth, may all feature prominently in our judgments for, as we have already seen, the relations between theories and things are far from simple. But testing, or as Quine more figuratively puts it, facing 'the tribunal of sense experience', is none the less important, whatever the failings of doctrinaire empiricism. Though much of our present understanding of what is involved in assessing a theory's relation to evidence derives from the empiricist tradition, neither that understanding, nor the philosophy in which it is rooted, is as monistic as some critics might have us believe.

In fact the philosopher who has been central to so much discussion of these questions in both philosophy of science and sociology, Karl Popper, is hardly an empiricist. Though his emphasis on constantly meeting our theoretical conjectures with potential refutations shows a healthy respect for the empirical, it is rooted in a thorough critique of the traditional empiricist concept of verification. Popper has, of course, been misread. I have heard more than one respected social scientist offer an account of Popperian philosophy in 'dogmatic falsificationist' terms, a vulgarisation which parallels that so often applied to Hempel. (1) But it is not my intention here to right such wrongs; Popper has not been short of defenders and interpreters. What I do want to suggest is that exploration of the Popperian tradition is essential for any discussion of demonstration.

Though Popper has been primarily concerned to formulate rational criteria for the appraisal of theories that are 'internal' in the sense that they transcend specific socio-cultural contexts, and though his philosophy of science is essentially normative, he did establish the parameters of much modern discussion. Accordingly, this chapter will revolve around the Popperian perspective, or, at least, certain aspects of it. It is inevitable that in focusing discussion in this way I shall be omitting views that would be of significance in any full discussion of demonstration. That cannot be helped. Arguments about the relation between theory and evidence have a way of becoming arguments about the whole structure of scientific inquiry. I cannot wholly avoid that escalation here, but I shall try to restrict it.

CONJECTURES AND REFUTATIONS

There are versions of Popper's philosophy (often at second and third hand) in which a few of his arguments are cut loose from the corpus of his work and herded into a pen marked 'falsification'. Thus isolated, they can be used to represent him as the author of a nice logical point about the atomistic relationship between an hypothesis and the evidence which is its potential refutation. This Popper is a fiction, a straw man erected either to be laid low or to be quoted in justification of indefensible positions. As he once observed: 'Criticism of my alleged views was widespread and highly successful. I have yet to meet a criticism of my views.' (2) Though he could hardly say the same today (there are cogent and stimulating criticisms of Popper) it is true that all too many accounts of his position are, at the very least, misleading.

Some part of that is a consequence of change on Popper's part. He began serious consideration of the issues that were to dominate his work in, by his own account, 1919, and it would be surprising if there were a single Popperian analysis given consistent expression throughout all those years. Falsification may be the 'touchstone' of his philosophy of science, but the developments found in the transition from 'The Logic of Scientific Discovery' to 'Conjectures and Refutations' and from there to 'Objective Knowledge' are hardly insignificant. (3) I do not intend to chart these changes, and nor do I have any ambitions toward comprehensive exposition. Instead I shall return to the analysis presented in 'The Logic of Scientific Discovery', filled out a little by reference to the later work, in the hope that this will serve to focus some of the problems involved in the idea of 'testing'. That may not be too misleading. Amsterdamski has suggested that the core Popperian position is best understood as 'naive falsificationism', although 'Conjectures and Refutations' does indeed suggest some modifications. The terminology, of course, comes from Lakatos, who has made a similar observation in his own inimitable fashion: 'the real Popper consists of $Popper_1$ together with some elements of $Popper_2$.' (4) $Popper_1$ is the 'naive' version; $Popper_2$ the sophisticate. To return, then, to the central Popperian text is not entirely irresponsible.

In so doing one point has to be made before any other: Popper's account of the ways in which scientific theories are to be assessed is inextricably bound up with what he terms the 'problem of demarcation'. Specifically, the need for criteria which will distinguish between the empirical

sciences proper and a ragbag of other pursuits including
metaphysics, mathematics, logic, and 'pseudo-science'.
That difficulty, of course, is hardly new; as Popper himself observes Kant had made it into the central problem of
epistemology. But for Popper it took on particular force
because of the intellectual environment in which he found
himself at the end of the Great War. Four current theories
dominated his concerns: Einstein's relativity, Marx's
theory of history, Freud's psychoanalysis, and Adler's
individual psychology. In what ways were they scientific?
Or, more important, what was the source of Popper's growing conviction that the one (Einstein) was indeed an instance
of empirical science, while the others were not? 'I felt
that these other three theories, though posing as sciences,
had in fact more in common with primitive myths than with
science; that they resembled astrology rather than astronomy.' (5) But how was this boundary to be drawn?

Traditionally, inductive method had been thought to be the
distinguishing characteristic of science. This 'scientific
method', it was argued, enabled us to make valid inferences
from singular statements, such as observations, to the more
universal statements characteristic of science. In practice this meant recording our experience in a systematic
way and, on the basis of this process of methodologised
observation, arriving at general scientific conclusions.
To Popper this view was fundamentally indefensible.
Partly, of course, because observation itself is theory-laden; one requires a point of view in order to choose
among the multiplicity of possible singular statements. But
more important to Popper was Hume's argument that there is

no logical justification for inductive inference, no reason why we should accept that further instances as yet unexperienced will resemble those already experienced. In what has become one of the cliches of the literature: however many white swans we observe we will not be justified in concluding that all swans are white.

Popper dubs the problem of induction 'Hume's problem' and, in effect, seeks to show that it is a facet of the more general problem of demarcation ('Kant's problem').

Why, I asked, do so many scientists believe in induction? I found they did so because they believed natural science to be characterised by the inductive method – by a method starting from, and relying upon, long sequences of observations and experiments. They believed that the difference between genuine science and metaphysical or pseudo-scientific speculation depended solely upon whether or not the inductive method was employed. They believed (to put it in my own terminology) that only the inductive method could provide a satisfactory <u>criterion of demarcation</u>. (6)

Clearly the question of induction raises more complex issues than are touched upon here. For present purposes, however, it is sufficient to recognise the force of Popper's rejection of a <u>logically justifiable</u> process of inductive inference, a rejection that led him to seek an alternative solution to Hume's and Kant's problems. The clues to that alternative were already to be found in his uneasy response to Marx, Freud, and Adler. What bothered him about these theories was that, once accepted, their proponents saw the world as replete in verifications of their views and, concomitantly, it was difficult to conceive of any way in which such theories could be shown to be wrong. There was Popper's criterion - 'could be shown to be wrong'. The distinguishing feature of empirical science, he argues, is that its statements admit of contradiction. They are, at least in principle, falsifiable. If we seek only for confirmations then they are easy enough to find. Genuinely to test a theory is not to seek for confirming instances but to seek to falsify it.

Not that falsification exhausts the process of testing. Popper also mentions criteria of internal consistency, logical form, and comparison with other theories in terms of scientific advance. (7) Yet it is the main focus of his attention, the central pivot of his solution to the problem of demarcation. Falsification is the guarantor of scientificity: 'it must be possible for an empirical scientific system to be refuted by experience.' (8) It is only in this possibility that we can logically distinguish science from pseudoscience. By the classic logic of the modus tollens we are permitted to argue from the <u>truth</u> of a singular statement to the <u>falsity</u> of a universal one. Take (to use Popper's own nomenclature) a system, t, of which p is a conclusion. The relation between t and p is one of logical implication and may be expressed thus: $t \rightarrow p$. If p is shown to be false ($\bar{p}$) then, since p derives logically from t, t is also false. Expressed symbolically: $((t \rightarrow p).)\bar{p} \rightarrow \bar{t}$. Note, however, that what is falsified by showing $\bar{p}$ is the whole system t, including whatever conditions may have been necessary to establish the relation of implication in the first place. So although falsification is logically defensible, there remains the problem of identifying quite what has been falsified.

What, then, has Popper achieved in introducing the criterion of falsification? He has, first of all, provided a logically defensible solution to the problem of demarcation. That is not to say that science is in fact distinguished by its reliance on falsification. But the criterion that he has offered, unlike that of the inductivists, is logically acceptable: we can indeed use singular statements to falsify universal ones. Second, he has sidestepped (rather than resolved) the problem of induction, at least as it was conventionally raised. Third, he has pinned down the difficulty at the heart of his mistrust of work like that of Marx's or Freud's. Their theories, in Popper's view, are too often irrefutable. They offer world-views to which one may be converted rather than systems of scientific theory. And lastly he has laid the foundations for an account of scientific demonstrability and scientific development based in the idea of meeting bold conjectures with strict refutations.

Now there are some versions of Popperian falsificationism that would stop here. All that is scientifically significant, they would suggest, is falsification. Science develops because we continually reject inadequate theories on the basis of factual evidence. Such 'dogmatic falsificationism' is problematic in ways that I shall not discuss; the view is forcibly dealt with in the literature. (9) What should be stressed, however, is that Popper himself is much concerned with the problems that arise from a falsificationist emphasis. Though falsification may be the key to Popper's analysis of science, he is not the dogmatist sometimes portrayed by unwary expositors. That is why it is important always to recall the centrality of the demarcation problem.

Falsification is first and foremost a resolution of Popper's difficulties in distinguishing science from non-science; it doesn't pretend to be an exhaustive account of the nature of demonstrability. Indeed, had he started with a question other than demarcation, the idea of falsification might not have assumed the prominence that it did. It is embedded in a larger discursive context, and it is in that context that it must be understood.

One might observe, for instance, that Popper is quite clear that the process of testing is an inter-subjective process and that any claims to 'objectivity' must be understood within that framework. In turn that view relates to our conceptions of 'truth' and to a range of epistemological problems that ought to be explored vis-a-vis the (apparently) changing views found in 'Objective Knowledge'. (10) To do so, however, would take me a good way from the topic of demonstration with which this chapter is concerned, however fair it might be to Popper. Accordingly, I shall attend to the more general issues raised in Popper's discussion only inasmuch as they are germane to my immediate problem.

In practice this means looking at Popper's attitude to conventionalism and the so-called 'Duhem problem', at some difficulties raised in the crucial area of what Popper calls 'basic statements', and at some general features of Popper's analysis of corroboration. All these arise directly from the central falsificationist argument. The 'Duhem problem' because of difficulties in deciding what is falsified by a specific test; 'basic statements' because they constitute the evidence for testing; and corroboration because it is the central concept in Popper's account of acceptance and growth of scientific knowledge. These three topics, then, form the core of the next section.

CONVENTIONS, BASIC STATEMENTS, AND CORROBORATION

First, then, conventionalism, and the associated claim that 'it is always possible to find some way of evading falsification, for example by introducing <u>ad hoc</u> an auxiliary hypothesis, or by changing <u>ad hoc</u> a definition.' (11) To Popper that criticism is just – it is possible to evade falsification – and the conventionalist view from which it stems 'is self-contained and defensible.' 'Attempts to detect inconsistencies in it', he goes on, 'are not likely to succeed.' (12) Yet he rejects conventionalism as a doctrine in that it advances a view of the fundamental goals and practices of science which is quite at odds with his own. Accordingly, his response to the claim that one can always evade falsification by appropriate adjustments to the system is to suggest that 'empirical method' should legislate against such 'conventionalist strategems'. Methodological rules, themselves conventions, must be introduced in order to avoid conventionalist traps. In effect, that is, he reasserts the primacy of falsification by proposing the introduction of a methodology which will facilitate that falsification. Thus the confrontation between Popperian falsificationism and conventionalism is precisely that: a confrontation. It is less a case argued from one side or the other, more two differently conventionalist views of science glaring at each other across a no-man's-land.

To adopt conventionalist strategems, then, is to lower or

destroy the scientific status of a theory. The Marxist theory of history, Popper argues, when faced with falsifications of its predictions, was reinterpreted in order to evade falsification. But in so doing Marx's followers made the theory irrefutable so undermining any claim it had to scientific status. (13) Whether that is an appropriate criticism of Marxist theories is unimportant. Its significance here is that it illustrates how 'hard' a criterion Popper is trying to develop. Having admitted that conventionalism is 'self-contained and defensible' and that its strictures on simple falsificationism must be taken seriously, Popper himself uses a conventionalist strategy to rescue falsificationism. To introduce methodological rules which spell out the circumstances under which falsification may be said to operate is to claim that, by convention, scientists decide the circumstances of falsification. (14) To that degree, then, Popper himself is a conventionalist, but a conventionalist with a prior commitment to the view that science must be distinguished from non-science in terms of its critical rationality. Once given that commitment, and given its specific expression in the principle of falsifiability, the rest is a question of arranging an appropriate methodology.

But is that a good defence against conventionalist criticism? To begin to make a judgment on that one must first recognise that there are different degrees of conventionalism. That normally associated with Duhem, for instance, is not as strong a claim as that more recently associated with Quine. Making a rough and ready distinction one could say that Duhem's case against falsificationism is that we have no reason to suppose that a particular falsification

applies to a particular hypothesis. As Popper observes in his discussion of the modus tollens it is the whole system, t, that is falsified if we show that p does not hold. (15) Duhem puts it this way:
> When certain consequences of a theory are struck by experimental contradiction, we learn that this theory should be modified but we are not told by the experiment what must be changed. It leaves to the physicist the task of finding out the weak spot that impairs the whole system. No absolute principle directs this inquiry, which different physicists may conduct in very different ways without having the right to accuse one another of illogicality. (16)

In such circumstances, as Popper suggests, 'it is sheer guesswork which of its [the theoretical system's] ingredients should be held responsible for any falsification.' (17) He goes on to suggest that this is of no great import to one who believes that all theories are guesses anyway (what's another conjecture here or there?), but surely that avoids the issue. In Popper's account theories may indeed be guesses, but they are guesses given credibility (and, later, corroboration) in the process of falsification. If that process itself is founded upon 'guesswork' as to which part of the system has actually been refuted, then the firm foundation of defence against conventionalism is no longer firm.

While not distinguishing between the Duhem and Quine elements of the thesis, Sandra Harding puts it neatly:
> Does the fact that in some cases scientists reach intersubjective agreement, at least temporarily, as to which part of their theories to revise save Popper's falsificationism from the Duhem-Quine thesis? The latter would appear to

> challenge not this uncontroversial sociological fact but the notion that it is tests which determine <u>which</u> part of our web of hypotheses and beliefs should be counted as refuted. (18)

Although falsifiability-in-principle may provide us with a criterion of demarcation, in practice it does not give us rational and conclusive grounds for refuting any particular element of a theory. Perhaps that doesn't matter? Perhaps demarcation is really the central problem and the rest merely a matter of methodological convention? There are certainly elements of that in Popper's argument. But if that is the case, what is to stop the slide from Duhem to Quine? Once we recognise that falsification does not allow us to distinguish which precise feature of the system (background knowledge, auxiliary hypotheses, operational procedures) is faulty, why should we not argue, with Quine, that 'any statement can be held true come what may, if we make drastic enough adjustments elsewhere in the system'? (19)

And if we can argue that, if we can claim as Quine does that 'our statements about the external world face the tribunal of sense experience not individually but only as a corporate body', then where is the force in Popper's claim that conventionalist strategems must be met with appropriate methodological strictures? (20) Why should we seek such restrictions if they run against the character of scientific knowledge as we understand it? In the end the only possible answer is that offered by Popper himself in his discussion of demarcation in 'The Logic of Scientific Discovery':

> My criterion of demarcation will accordingly have to be regarded as a <u>proposal for an agreement or convention</u>.

> ... Thus I freely admit that in arriving at my proposals I have been guided, in the last analysis, by value judgments and predilections. (21)

Quoting this passage is not intended to be a criticism of Popper. Rather I want to emphasise that the basic division between falsificationism and conventionalism is that between commitment to a critical rationalist prescription for scientific methodology and a view of science that wholly recognises the constitutive role played by conventional practices.
From the falsificationist point of view the choice is clear. As Lakatos puts it: 'one has to choose between some sort of methodological falsificationism and irrationalism.' (22) But what if one accepts neither an all pervasive rationalist commitment nor the view that the alternative is simply irrationalism?

I shall return to that theme later. For the moment let us continue with Popper on the assumption that we accept his case against conventionalist strategems. That given, and if falsification is actually to be practicable, we must define the empirical grounds on which our refutations are to be based. If falsification involves passing from the truth of singular statements to the falsity of universals, then on what basis do we establish the truth of singular statements? Popper raises that question in terms of 'basic statements' — 'a statement which can serve as a premise in an empirical falsification; in brief a statement of singular fact' (23) — and gives the matter extended discussion throughout 'The Logic of Scientific Discovery'. Rather than closely following the course of his exposition (which is scattered through his text) I shall try to reconstruct his basic argument. (24)

Once we have legislated against resort to conventionalist stratagems, falsifiability itself is to be understood in terms of the logical relation between theories and basic statements. Care must be taken here. Basic statements are not 'basic' in the sense that they are ultimate truths. Popper makes it clear that science has no ultimate statements in as much as there are no statements in science which cannot, in principle, be refuted. Indeed, 'basic statements are accepted as the result of a decision or agreement; and to that extent they are conventions.' (25) In these terms a theory is said to be falsified if we accept basic statements that contradict it. Not all basic statements, then, are necessarily <u>accepted</u>. In Popper's view the whole system of basic statements includes all self-consistent singular statements of the appropriate logical form; some will be permitted by the theory while some will be inconsistent with it. Falsification turns on which basic statements we actually accept, and the central questions become: what are the crucial features of basic statements and on what grounds are we to accept them in the course of falsification?

Are they, for example, 'objective' statements of fact? Given the general thrust of Popper's argument that could hardly be the case. He is clear that accepting a basic statement is a result of methodological decisions. Thus, although the system of basic statements includes all conceivable singular statements of fact, their 'objectivity' resides in their inter-subjective testability, and not in any absolute characteristics of basic statements qua basic statements. But if this is the case, and if none the less we wish to preserve a meaningful sense in which our theories are tested in relation to an empirical domain, then in what ways do basic statements represent that domain? How, for example, do they relate to our perceptual experience? Popper approaches this problem by introducing 'Fries trilemma'. (26)

Fries, he says, refused to accept that scientific statements should be asserted <u>dogmatically</u>. Rather they must be <u>rationally justified</u> in some way, and if that is the case then we must recognise that, logically, statements can only be justified by other statements. We can construct an argument in justification of a particular statement but that argument will be based upon and expressed in yet other statements. Inevitably, then, if we demand the justification of all scientific statements we enter upon an infinite regress: each 'justification' gives us yet more statements which must be justified. How are we to resolve this dilemma? Either we accept that at least some scientific statements must be dogmatically asserted, or we are caught in a regress in which there is no foundation for scientific knowledge. Fries's resolution (and, according to Popper, that of most other epistemologists) is to resort to forms of <u>psychologism</u>; to argue that our direct sense experiences enable us to justify scientific statements. This view, though, is problematic for precisely those reasons Popper has already advanced in relation to inductive inference: 'we can utter no scientific statement that does not go far beyond what can be known with certainty "on the basis of immediate experience".' (27) What is more, although Fries's discussion dates from the first half of the nineteenth century, the problem of psychologism (to Popper at least) is still to be found in more recent empiricist theories.

Predictably, Popper's solution is akin to his response to conventionalism. Just as one deals with conventionalist strategems by methodological decisions, so, in testing a theory, one stops at basic statements which one decides to accept. Basic statements, that is, are matters for intersubjective agreement – though undogmatic in that they can be tested further. Clearly, as Popper himself observes, this will lead to a regress, but a harmless regress in that no process of proof rests upon it. But what of the dread beast psychologism?

> I admit, again, that the decision to accept a basic statement, and to be satisfied with it, is causally connected with our experiences – especially with our <u>perceptual experiences</u>. But we do not attempt to <u>justify</u> basic statements by those experiences. Experiences can <u>motivate a decision</u>, and hence an acceptance or a rejection of a statement, but a basic statement cannot be <u>justified</u> by them – no more than by thumping the table. (28) (Popper's italics)

If this is the case, if the 'empirical basis' is grounded in decisions rather than in some absolute relation between perceptual experience and individual statement, then what are the crucial characteristics of basic statements? They are not justified by perception or any similar 'psychologistic' link. Yet they are 'singular existential statements' asserting 'that an observable event is occurring in an individual region of space and time.' (29)

'An observable event.' Does not this reintroduce the psychologism that Popper is so eager to avoid? In what sense can we speak of an observable event without making reference to perceptual experience? Popper's claim is that although observations and perceptions are psychological, to attribute observability is not. To demand of basic statements that they deal with 'observable events' is, Popper suggests, to demand that they deal in events 'involving position and movement of macroscopic physical bodies', a non-psychologistic sense of the term 'observable'. (30) But it is also an unacceptably restricted sense of 'observable event'. Science, even natural science, admits of broader ideas of observability (depending on the specific auxiliary theories with which it operates) and if such a restriction upon observability is a product of Popper's desire to avoid psychologism then it seems reasonable to wonder if psychologism would not be preferable. Indeed, considering how important the concept of observability appears to be to Popper's account of basic statements, his comments in 'The Logic of Scientific Discovery' are somewhat brief and evasive:

> I have no intention of defining the term 'observable' or 'observable event', though I am quite ready to elucidate it by means of either psychologistic or mechanistic examples. I think that it should be introduced as an undefined term which becomes sufficiently precise in use: as a primitive concept whose use the epistemologist has to learn. (31)

Up to a point, however, the issue of observability is something of a by-way in Popper's methodology. His central emphasis is on accepting or rejecting basic statements as decisions. There are no fixed external standards (such as an absolute criterion of objective observation) only intersubjective processes of methodological decision-making. In that sense, as Popper observes, basic statements are conventions and decisions upon them depend on rule-governed procedures. They have no features as such that warrant their use. Wherever we look - at 'objectivity', at logical form, at 'observability' - we find necessary but not sufficient conditions for recognising (though not accepting) a basic statement. And, of course, in Popper's world that is how it should be, for it is a first tenet of falsificationism that no statement can be absolutely warranted. Everything is conjectural and can therefore be refuted. It is in the methodological rules that we should seek to understand basic statements, for it is by virtue of 'a procedure governed by rules' that we agree to accept them.

Rules, then, are at the heart of the matter, and on their nature Popper is hardly specific. He singles out one which he considers to be of special importance. We shouldn't accept 'stray basic statements', he says, rather 'we should accept basic statements in the course of testing theories.' And later: 'agreement upon the acceptance or rejection of basic statements is reached, as a rule, on the occasion of applying a theory; the agreement, in fact, is part of an application which puts the theory to the test.' (32) What he is emphasising here is the theory-dependence of observation statements and the impossibility of maintaining (as did the positivists) that such statements are justifiable in our immediate experience. Falsification, then, is not a question of testing isolated hypotheses against unquestionable statements of fact. Instead it involves judging competing theories against conventionally accepted basic statements. 'From a logical point of view, the testing of a theory depends upon basic statements whose acceptance or rejection, in its turn, depends upon our decisions. Thus it is <u>decisions</u> which settle the fate of theories.' We decide, for the time being, that specific basic statements are sufficiently credible for us to arbitrate between the various theories that we develop. 'We choose the theory which best holds its own in competition with other theories.' (33)

How, then, can we make that judgment? Assuming agreement on which basic statements to accept, on what grounds do we choose among theories? There is now a sense in which falsifiability is a matter of degree (we are choosing <u>among</u> theories, not making a simple yes or no judgment) and <u>corroboration</u> becomes important. Our aim is not simply to falsify; we actually accept theories which have stood up to severe tests and are thus rendered more credible. But what is a 'severe test'? On what grounds are we to say that one theory is more corroborated than another? Popper's answers to such questions are lengthy and, at times, more than a little complex. Here I want simply to indicate the direction of his argument.

Commencing with the recognition that falsifiability is a matter of degree, he argues that, in choosing among theories, we must aim for the maximally falsifiable theory. The more potential falsifiers there are for a given theory, then the more that theory says about the world. This is so in the following sense. Such a theory rules out more basic statements than would one with less potential falsifiers, and so, Popper suggests, the information it carries, its empirical content, increases with its degree of falsifiability.

Care is needed here for this sense of 'empirical content' runs counter to common usage. To say that one theory has more 'empirical content' than another intuitively suggests that it incorporates a wider range of accepted statements. But this is not Popper's sense of the term, and a moment's reflection will suggest why. If we are unable to <u>verify</u> statements then the fact that a theory permits a large number tells us nothing about the viability of that theory. We are able to falsify, however, in which case a theory has more to say in as much as it rules out a larger class of basic statements. The more chance there is of falsifying a theory then the more that theory will tell us. Popper proposes a number of ways in which we might try formally to compare degrees of falsifiability, most notably in terms of 'logical probability' or, as he later terms it, 'absolute logical probability'. (34) And this, in turn, leads to appraisals of corroboration.

The whole foundation of Popper's argument, of course, is that theories are not verifiable in the sense assumed by the empiricist tradition. They can only be falsified. However, he does accept that theories may be corroborated; we can make judgments about how well a theory stands up to tests and so appraise it in terms of its degree of corroboration. This is not, as Popper is quick to point out, a question of the <u>probability</u> of a hypothesis. What is at issue, rather, is the severity of the tests to which a theory has been subjected. We cannot simply argue that the more times a theory has been tested and not falsified, then the more corroborated it is; that might come about because a theory precludes very few basic statements and is therefore faced with less potential falsifiers.

> It is not so much the number of corroborating instances which determine the degree of corroboration as the <u>severity of the various tests</u> to which the hypothesis in question can be, and has been, subjected. But the severity of the tests, in its turn, depends upon the <u>degree of testability</u>, and thus upon the simplicity of the hypothesis: the hypothesis which is falsifiable in a higher degree, or the simpler hypothesis, is also the one which is corroborable in a higher degree. (35)

Furthermore, a judgment of corroboration is a relativised judgment: it relates to a particular time-period and set of basic statements. We are not, then, saying anything about the <u>truth</u> of the corroborated statement; rather that the statement 'is corroborated with respect to some system of basic statements — a system accepted up to a particular point in time.' (36)

This is not the end of Popper's discussion, though it is as far as I shall take this exposition. Using Tarski's analysis of 'truth' he goes on to develop an account of how falsificationism can grapple with scientific progress and growth. (37) We will encounter some features of that discussion in relation to Lakatos's work later but, for the present, the more limited problem of demonstration must remain the focus.

FROM CORROBORATION TO INCOMMENSURABILITY

It is now possible to stand back a little from the Popperian enterprise. Though it begins with the problem of demarcation and the falsificationist argument, ultimately it becomes an attempt to formulate and prescribe a set of methodological rules which, because of their critical rationality, are presumed to encourage scientific growth. Popper and his followers are convinced of the supremacy of rationalism. Hence their deep antagonism to Feyerabend's project: 'to show that rationality is one tradition among many rather than a standard to which traditions must conform.' (38) What's more, Popper's is a particular version of rationalism, its development thoroughly coloured by his starting point in demarcation. That concern to draw an unambiguous line between science and non-science moulds all his subsequent thought. When he does offer an analysis of scientific growth it is an analysis overshadowed by his prior commitment to falsificationism. Though other rationalist accounts are possible (that offered by Lakatos, for example) Popper's basic choices have been made in the course of exploring demarcation. Hence the idea of corroboration, central to his understanding of the way we choose among competing theories, is in effect a by-product of his primary analysis of falsification. To claim that a theory has been corroborated is to claim that it has survived severe testing without being falsified. Accordingly, whatever is problematic about Popperian falsificationism is also problematic for his understanding of scientific growth. Corroboration, in this sense, is the other side of the coin of falsification.

Consider in this context, then, two of the problems that we have already encountered: that raised by conventionalism and that concerning the empirical status of basic statements. At the risk of being repetitious we can distinguish two lines of conventionalist attack. The first (associated with Duhem) revolves around identifying what precisely has been falsified in the course of a particular experiment. Do we condemn the hypothesis itself, or ascribe the failure to 'background knowledge' taken for granted in the course of the test? The second (associated with Quine) is stronger. It suggests that not only do we not know what has been falsified but, should we wish, we can save any hypothesis from falsification by making adjustments elsewhere in the system to which the hypothesis belongs. Popper's response to these sorts of claims, as we have seen, is to argue for methodological rules ('conventions') which will preclude resort to hypothesis-saving strategems and which will specify precise criteria for deciding when an hypothesis has been falsified. But even presuming that we can construct such a set of rules (and Popper surely does not do so) the evidence suggests that science does not actually operate in this way. Naturally, as a normative theorist, Popper would argue that we need such rules if scientific progress is to be ensured: only by observing the canons of critical rationality will science grow. Yet the recently influential work of such historians and philosophers of science as Kuhn and Feyerabend suggests that science has often progressed precisely by resort to those activities Popper would like to proscribe. (39) So unless we are absolutely committed to critical rationality, it is by no means clear that falsificationism can be defended against conventionalist doubts. And if that is the case then those doubts must also apply to corroboration.

If, in the last analysis, we cannot tell what precisely has been falsified, then how can we tell whether or not a particular hypothesis has undergone 'severe' test? And, if that cannot be established, then we can hardly make the appraisals of corroboration so central to Popper's conception of scientific growth. Furthermore, if we can, as Quine suggests, save any hypothesis from falsification by making adjustments elsewhere in the system, then there must also be a sense in which we can provide corroboration at will, a patently ridiculous notion from a Popperian point of view. Let me be clear about what I am arguing here. Naive falsificationism has a problem in conceptualising scientific progress: if our endeavour is constantly to seek to falsify, how then can we accumulate knowledge? Popper, in common with other 'methodological falsificationists', deals with that problem by developing an analysis of corroboration. We can choose to retain certain theories on the basis of an appraisal of their degree of corroboration. But, it would appear, corroboration is just as open to conventionalist doubts as is falsification; to make either work we require a single overarching set of strict methodological rules. Science, however, does not appear to operate with any such single methodological system, so the justification for its development must be in terms of a transcendent rationality. Popper's strategy is to take as axiomatic the view that science-proper is rationally conducted, and then try to squeeze a somewhat less than rational enterprise into this mould. In trying to grapple with the fact that science is not practised as restrictively as Popper might like, conventionalists and other critics seem to offer a more realistic understanding of scientific progress.

This is not merely a question of recognising that scientific 'success' cannot be associated with only one methodological system. Quine's argument suggests that decisions about the demonstrability of a theory are often taken at a more holistic level than that envisioned in the Popperian analysis. Popper is not too clear about the level(s) on which corroboration takes place. Do we speak of individual statements having been corroborated, or does corroborative status flow upwards from statements to theories? Much of Popper's analysis appears to apply to the <u>sentence system</u> as the locus of demonstrability without recognising that we also consider <u>models</u> to be demonstrable. That is, we do not simply subject isolated hypotheses to test and thus appraise their corroborative status. We also appraise our models in terms of their general 'fit', and we are unlikely to reject a favoured model because a hypothesis considered consistent with it has been falsified. Similarly, accepting a particular model as currently the best available involves something more than establishing corroboration on the basis of severe test.

If we now add in to this discussion the doubts arising from Popper's account of basic statements, then the whole question of corroboration begins to look increasingly difficult. Even without pursuing Popper into the areas of verisimilitude and corroborated excess content, it should be apparent that 'severe test' in relation to accepted basic statements is being asked to do a great deal of work. It is on these grounds that we make the all-important choice among competing theories. Popper emphasises that at all stages this is a matter of methodological decision, which means that his analysis is only fully meaningful if the 'rules' governing such decisions are available to us. Without that information there

is an important sense in which his analysis is empty.
Some of Popper's critics - particularly those who feel that he has not done justice to the question of induction - have even been inclined to argue that if Popperian corroboration has any substance it must involve induction. (40) If 'standing up to severe test' is to count for anything, they are arguing, then it must be a proposal to accept a theory (however provisionally) in relation to basic statements, so leading to a process of 'justification' on the basis of particular confirming instances. And if there is any force to that claim it offers us the irony of a confirmed opponent of inductivism forced to have covert recourse to inductive method.

I have no wish to make a judgment on such arguments. Here it is only necessary that we appreciate the tensions in Popper's work that encourage these diverse criticisms. The difficulty is that having recognised that scientific statements are fallible, that there is no final foundation for science, Popper seeks to avoid a thorough-going scepticism. Since there are senses in which science does work and grow, and since complete fallibilism cannot tell us why, Popper requires an additional analysis to grapple with scientific growth. This he develops by positing a methodology based in critical rationalism. We make rational <u>decisions</u> about falsification, about what should be treated as unproblematic background knowledge, and about accepting basic statements. But who is really to say that one decision is to be preferred to another, that one system of rules is distinctively scientific?

Lakatos - who has developed the most interesting extension of Popper's analysis - puts it this way:

If scientific theories are neither provable, nor probabilifiable, nor disprovable, then the sceptics seem to be finally right: science is no more than vain speculation and there is no such thing as progress in scientific knowledge. Can we still oppose scepticism? <u>Can we save scientific criticism from fallibilism</u>? (41)

On his account Popper's 'methodological falsificationism' is the most promising of such attempts (particularly in its 'sophisticated' version), and his own Methodology of Research Programmes is built upon this foundation. Let us then look to Lakatos for solutions to the problems left unresolved in Popper's analysis, beginning, unavoidably, by looking at Popper through Lakatos's eyes.

Popper's methodological falsificationism, Lakatos argues, is a form of revolutionary conventionalism. It recognises that all scientific statements are equally fallible, but allows that some <u>singular statements</u> may be made unfalsifiable by <u>fiat</u>. These are basic statements, and their (temporary) unfalsifiability is rooted in inter-subjective decisions. In making such decisions - for example, in utilising experimental techniques - scientists have recourse to certain other theories and they, too, are fallible. Under methodological falsificationism such theories are regarded in a given context as 'unproblematic background knowledge': 'the methodological falsificationist uses our most successful theories as extensions of our senses.' (42) Accordingly, methodological decisions are also necessary to demarcate that which is unproblematic background knowledge from that which is theory under test, decisions made particularly difficult where the question is whether to consider a ceteris paribus clause as part of the 'unproblematic' domain.

All this is familiar enough, if given a rather more conventionalist slant than is often found in commentaries upon Popper. The difficulty, Lakatos suggests, is that the image of science thus developed – though superior to most of its predecessors – is still dissonant with the history of science. It is not the case that a test is (or even ought to be) a direct confrontation between theory and data, nor is the only significant outcome a falsification. Tests are historically more complex than that, and confirmations are often significant. To resolve these inconsistencies between reconstruction and actual practice Lakatos recommends a shift from 'naive' falsificationism to a more sophisticated variant. He detects at least part of this development in Popper, drawing particularly on the idea that we accept a theory if it has 'corroborated excess empirical content over its predecessor (or rival), that is, only if it leads to the discovery of novel facts.' (43) This leads him to a conceptualisation in terms of a series of theories which may be 'progressive' (if each has corroborated excess empirical content over the one before) or 'degenerative' (if they do not). Appraisal is now directed at a whole set of theories; we are no longer dealing in isolated tests and falsifications.

Extended in this way, of course, the Popperian view grapples more successfully with the Duhem–Quine problems. Faced with a falsification the Lakatosian scientist doesn't have to decide which part of the 'theory' has actually been falsified. Instead, he or she can try replacing any part in the hope of arriving at an alternative theory which explains the initial anomaly 'progressively'. Until such a replacement is found the 'falsified' theory survives, and the question which bedevils naive falsificationism (which bit is false?) ceases to be important. Instead we seek alternatives that are corroborated and that incorporate the material contained in the initial theory. This strategy also offers a response to the stronger version of the Duhem-Quine critique: the view that a theory may always be preserved by altering other elements in the system. From Lakatos's point of view such defences are only permitted if they are part of a progressive programme. Thus (in an example provided by Worral) both Newtonianism and Marxism had recourse to conventionalist defences: in response to refutation their proponents argued that only certain auxiliary assumptions were at fault, while their main theories remained untouched. In the terms in which Popper discusses such 'conventionalist strategems', of course, the Newtonian and Marxist use would be indistinguishable. In Lakatos's terms, however, a distinction can be made by recourse to corroborated excess empirical content. The one research programme (Newtonianism) is demonstrably progressive, while Marxism - in as much as its defensive amendments do not involve excess content other than that which presented the problem in the first place - is not. (44) This is no place to judge the specifics of that claim. What is important is that Lakatos provides in principle for a way of distinguishing acceptable and unacceptable conventionalism.

Lakatos is offering us, then, an alternative to Popper's deductivism, a 'pluralistic model' in which the empirical basis is somewhat differently constituted.

In the pluralistic model the clash is not 'between theories and facts' but between two high-level theories: between an <u>interpretative theory</u> to provide the facts and an <u>explanatory theory</u> to explain them; and the interpretative theory may be on quite as high a level as the explanatory theory. (45)

And accordingly:

It is not that we propose a theory and Nature may shout NO; rather we propose a maze of theories, and Nature may shout INCONSISTENT. (46)

Demonstration, then, is not a question of a simple relation between theories and facts, but one of dealing with inconsistencies between theories. An experimentalist's interpretative theory is just as much up for replacement as that of a theoretician. Note, though, that this doesn't eliminate all the difficulties surrounding Popper's analysis of basic statements. We must still draw (conventional) lines as to what is to be counted as 'factual' for the purpose of the now all-important corroboration. In the 'methodology of research programmes' approach the crucial decision is that distinguishing progressive and degenerative programmes. If a theoretical series is progressive each element includes that which went before plus corroborated excess content, and corroboration reflects a relationship between a statement and the 'empirical basis'. Experience is still the court of appeal, even if it is experience mediated by conventionally accepted interpretative theories.

Lakatos builds this into an account of research programmes ('reminiscent of Kuhnian "normal science"') and their continuity. (47) Research programmes, he suggests, are characterised by sets of methodological rules or <u>heuristics</u>. In the <u>negative heuristic</u> certain activities are proscribed –

prominently, any attempt to falsify the <u>hard core</u> of the programme. Instead the <u>hard core</u> is to be preserved by surrounding it with a <u>protective belt</u> of auxiliary hypotheses which are subject to testing, reformulation, and replacement. If alterations to the protective belt lead to progressive problem shifts, that is to an increase in corroborated excess empirical content, then a research programme can be deemed successful. In the <u>positive heuristic</u>, by contrast, we are provided with suggestions as to the directions in which a programme may develop. The positive heuristic gives rise to 'a chain of ever more complicated <u>models</u> simulating reality: the scientist's attention is riveted on building his models following instructions which are laid down in the positive part of his programme. He ignores the <u>actual</u> counterexamples, the available "<u>data</u> ".' (48) Models, then, are created to be replaced; from their progression grows the progression of the whole research programme.

This sort of view, Lakatos claims, is far more a reflection of the history of science than is earlier falsificationism. Indeed, it is possible to see his work as an attempt to incorporate the insights of history of science into critical rational methodology; hence the resemblance between Kuhn's 'paradigm' and Lakatos's 'research programme'. But there is one issue, raised by Kuhn and pursued enthusiastically by Feyerabend, which remains unincorporated: incommensurability. Since the early 1960s Kuhn, Feyerabend, and others have repeatedly observed that 'paradigms' (or whatever concept different authors use) cannot be compared to one another because the terms in which they are cast, and the interpretations they offer of evidence, are fundamentally different. There is no trans-paradigmatic vantage point from which we can make considered evaluations of their relative merits. Paradigms are, so to speak, mutually untranslatable. And related to that is the claim that the transition from one paradigm to another may involve losses in knowledge as well as gains: the phenomenon of 'Kuhn loss' as it has been labelled. Thus a new paradigm doesn't always or necessarily incorporate everything of value established in the old one.

Such claims as these – that paradigms are incommensurable, that they cannot be compared in relation to evidence because there is no trans-paradigmatic source of such evidence, that the transition from one to another often leads to 'loss' of knowledge – are not restricted to Kuhnian paradigm theory. (49) They are pertinent to any account of science which presumes that we can make rational judgments as between theories, paradigms, or, indeed, research programmes. Thus if we accept the incommensurability case then severe restrictions must be placed on the 'rationalism' of even Lakatos's analysis. His model hinges crucially on our ability to choose among programmes, distinguishing those that are progressive from those that are degenerative. But how do we make that choice if the theories involved are incommensurable? For Lakatos the measure of progressiveness is corroborated excess empirical content: 'if each new theory leads us to the actual discovery of some new fact.' (50) But that assumes an empirical basis independent of the theories under judgment, one which can provide 'facts' that remain unchanged from one theory to the next. In some cases that may be possible. But if there are any incommensurable theories then, in those cases, the judgment cannot be made, and the claim of research programme analysis to universality would be unfounded. Lakatos, then, has recognised the force of contemporary criticisms of critical rationalism, and has tried to adapt Popperian analysis to the new philosophy of science. That is a considerable advance, but one which remains incomplete and – in its fundamental commitment to rationalism – perhaps indefensible. Scientific demonstration is more complex still than that.

ON DEMONSTRATION

Much of that complexity derives from those features of science that lie beyond the bounds of critical rationalism, some of them in the domain scathingly referred to by Lakatos as 'mob psychology'. Decisions that we take about the demonstrability of statements and theories depend upon many factors, only some of which can be adequately grasped or expressed in formal analysis. In the rationalist tradition there is a tendency to dismiss from science those practices not amenable to rationalist analysis: they are not part of science, the argument runs, or if they are they certainly shouldn't be. That is not surprising, of course, in a tradition which takes as its starting point the problem of demarcation. If you begin by assuming that science displays some unique quality that makes it science, then, inevitably, you will have to dismiss those features that do not fit into your analysis - a strategy consummately employed by Popper. But if we are ever fully to portray scientific demonstrability we will need more than that. A study of both the sociology and history of science, as well as the philosophy, is essential, brought together in a conceptualisation which doesn't reduce one to another but which allows us to explore their interaction. A restricted discussion such as this clearly falls far short of that ambition. However, we can make some tentative comments about some of the factors involved, if only to suggest points of departure.

Whatever his failings, though, it does seem that Popper's emphasis on <u>decisions</u> is appropriate. At all stages of our claims to 'demonstrate' scientific statements and theories we do indeed operate by making methodological decisions - decisions subject to all the pressures, rules, and practices of the scientific community. There is no reason to suppose (or, indeed, to recommend) that these decisions can be best understood in terms of a single rational methodology. As I have suggested demonstration is more complex than that, and the challenge is to grasp it in its complexity, not reduce it to neat and tidy formulae. It is conventional to talk of 'rules' here: 'rules' governing the processes by which science generates its data and arrives at its conclusions. I have used the term myself in chapter 2, qualifying it somewhat in chapter 4. As I argued there the term can be misleading, particularly if it suggests that scientific practice is carried on within a clear, unambiguous, and all-embracing methodology. We have methodologies, certainly, but they can no more be rendered down into formal prescriptive systems than can any other form of social activity.

If that is so, what <u>can</u> we say about demonstration in the sorts of terms in which this discussion is cast? Is philosophical analysis to be dissolved into sociology? I think not. Formal features of our reasoning are important and can be grasped in philosophy-of-science terms. On the basis of this discussion, for instance, it seems reasonable to suggest that demonstration is not exhausted by facing individual statements with factual evidence, neither in terms of providing confirmation nor with the intent to falsify. There are systems of scientific statements which are immune to falsification - for a considerable period of time, if not for all eternity. Some of those statements are <u>universal</u> statements; conventionally generated immunity is not restricted to <u>singular</u> statements as it is in Popper's account. Lakatos recognises that, of course, and tries to grapple with it in his idea of the unfalsifiable <u>hard core</u> and his reference to the importance of <u>models</u> in the positive heuristic. Whatever difficulties his approach encounters in distinguishing between progression and degeneration, this central imagery is persuasive. Missing from it, however, is a more holistic sense of 'demonstration'. Once we recognise that demonstration involves more than simply facing individual statements with 'facts' we encounter the possibility that, although some theories may be unfalsifiable by fiat, others may be considered to be demonstrated as a whole. The claim that a particular model is a good overall 'fit' to the empirical world is also a form of demonstration. That is, it involves facing our ideas with evidence, but pitched at a more holistic level and, presumably, employing somewhat different criteria to those we might invoke in relation to isolated statements.

What of the evidence, though, what can be said of that? Clearly it is a product of both the world that we are studying and the specific interpretative theories that we use to generate observations. The claim is familiar enough (the whole Popperian tradition recognises that 'observations' are theory-laden) but it has far-reaching consequences. In particular, as Lakatos observes, it shifts attention to the idea of consistency among theories and hence underlines the relativity of demonstration. Decisions about which interpretative theories are to be employed are not fixed and immutable (though they are often treated as if they were) and they are taken on a variety of grounds not all of which are 'rational' in the conventional sense. And that, in turn, suggests that science cannot be 'saved from fallibilism' as Popper tries to do. Science, like all human practices, is fallible, and to demand that it should be 'saved' is a misconception finally rooted in a commitment to the superiority of scientific rationalism. We should embrace fallibilism - not in the sense of abandoning all hope of method and rigour, but in recognising that science's very fallibility demands of us an additional clarity. Clarity about the various theories that we use to create 'empirical evidence'; clarity about the models that we accept and maintain whatever the 'facts'; and clarity about the grounds on which we accept theories as somehow corroborated. At root decisions on such topics are ideological, and it is possible that different ideologies may be incommensurable. But that would be no justification for our being unclear in our grasp of what we are doing when we make claims about the relations of theories to each other and to the world they putatively model.

In sociology ideas about demonstration were for many years bedevilled by crass empiricism. That is no longer the case (though empiricism has certainly not disappeared from sociological research) and there is increasing recognition that a central task for sociological work involves exploring the interpretative theories that sociologists use to generate 'data'. Work in phenomenology, ethnomethodology, and the like, has begun to make some progress, though crude claims about 'staying close to the phenomena' and thus ensuring better 'evidence' have tended merely to substitute one unquestioned interpretative theory for another. Philosophy of science does not offer sociology a solution to these problems, a prescription as to <u>how to do</u> demonstration. But it can help to direct our attention, help us to identify some of the problems to which we must attend if we are honestly to face our theories with 'the tribunal of sense experience'.

The broad argument of this chapter can be summarised thus:

1 Popper and his followers have been central to discussion of demonstrability in philosophy of science, and Popper's point of departure is the 'problem of demarcation'. His argument for falsificationism is an attempt to 'solve' this problem.

2 Three important areas of difficulty may be isolated:
   (a) problems of conventionalism
   (b) problems of basic statements
   (c) problems of corroboration

3 Popper's response to the conventionalist argument that 'falsified' hypotheses can always be saved is to argue in favour of a methodology that precludes resort to such 'conventionalist strategems'. This argument is based in his commitment to critical rationalism.

4 He argues that we accept certain basic statements as empirical grounds for falsification on the basis of methodological <u>decisions</u>, not psychologistic claims about observations.

5 Furthermore, he argues, we are concerned not simply with falsification; we also choose to accept one theory over and above another where it is better 'corroborated', i.e. it has survived severe test.

6 Through corroboration Popper hopes to understand scientific growth, but corroboration is open to similar doubts as is straightforward falsificationism.

7 As Lakatos suggests the problem is: 'Can we save scientific criticism from fallibilism'? He argues that we can by developing Popper's ideas into what Lakatos terms the 'methodology of research programmes'.

8 There remain, however, crucial difficulties about incommensurability and 'Kuhn loss'.

9 The problems encountered in the Popperian tradition illustrate that scientific practice cannot be understood without invoking factors outside the realm of critical rationalism. Demonstrability is indeed a complex process.

chapter 7

Methodological diversity in sociology

> An adequate, comprehensive political and social theory must be at once empirical, interpretative, and critical.
> Richard J. Bernstein, 'The Restructuring of Social and Political Theory'

Most of this book has been unremittingly abstract. Though I have employed some sociological examples, they have not been common; the weight of my attention has been elsewhere, with the debates and difficulties of twentieth-century philosophy of science. In this final chapter, however, I want to pay more attention to sociology. Schematically, still, and undoubtedly abstract, but at least focused on specific features of sociological inquiry. In chapter 1 I suggested that one view underlying this study was 'naturalism': the claim that there is a structure of inquiry common to all empirical disciplines, or, if you prefer, to all 'sciences'. I have not sought a priori to demonstrate that claim. (1) Rather I have tried to explore some of the ways in which traditional philosophy of science has proved problematic in application even to disciplines universally accepted as scientific, hence suggesting that the features often presumed by sociologists to distinguish science from other forms of inquiry do not, in fact, serve that purpose.

If this case can be maintained - and modern 'post-empiricist' philosophy of science does lend it some support - then it is reasonable to suppose that philosophy of science remains pertinent for sociologists. In particular, those interested in the evident variety of sociological procedure, in comparative methodology, can find topics of some significance addressed in the philosophy of science. Sociologists, like other scientists, encounter problems about the nature of their theories, about the bases on which choices may be made among competing views, about the nature of what can be counted as empirical materials, and about the sorts of accounts which constitute acceptable explanations. That is to say, sociologists, too, develop 'rules' which define acceptable methodological practice in different areas of the discipline. In this chapter I want to explore some of that variation by developing a general classification of types of sociological practice.

Such an endeavour is hardly new; there have been many attempts to classify sociologies, both simple and complex, polemical and analytical. The majority have been cast in substantive terms, their classifications reflecting the characteristic assumptions about subject matter found in different sociological perspectives. (2) My focus here is more formally methodological. Not because I believe that substantive questions are unimportant, nor even that they have no procedural implications. On the contrary, I do consider them to be of considerable significance and they surely have implications for methodological practice. That topic, however, is best reserved for fuller consideration elsewhere. Meanwhile, I am convinced that there is something to be learned from considering sociological procedures separate from the constitution of that discipline's subject matter. Clearly, then, I would not accept the sort of case typified by Winch's 'The Idea of a Social Science', in which it is argued that the nature of sociology's subject matter demands distinctively non-scientific methods. (3) In part that is because Winch's view of scientific practice seems to me to be inadequate. He assumes that science operates in certain ways (many of them central to traditional philosophy of science reconstructions) while much modern work, as we have seen, suggests that this is simply not the case. Science does not function as the received view suggested that it did. And, less specifically, it is surely more important to try to grasp sociology's undoubted methodological variation, than to make an a priori case that sociologists should behave in certain ways. Hence the general classification and analysis which follows.

DIMENSIONS OF VARIATION

The aim of the exercise, then, is to develop a simple typology. To that end I shall suggest two dimensions of variation, each conceptualising a different aspect of the relation presumed to exist between our 'scientific knowledge' and the 'empirical world'. Since this is such a fundamental methodological point of reference, the property space thus generated encompasses a number of other familiar conceptual oppositions. I shall mention some of them later, but first it is necessary to define the axes of my typology. One is essentially a distinction to be found embedded in the terms of reference of Figure 2.1 - particularly the contrast to be made between an emphasis on 'experience', on the one hand, and on 'demonstration' and 'explanation' on the other. It is a significant feature of many general methodological discussions in sociology that these complementary elements of inquiry, as I have conceived them, are given one-sided emphasis, some perspectives stressing the central importance of 'experience', while others over-emphasise the role of the empirical 'test'. In employing a similar distinction here I am not implying any such polemical commitment.

My contrast, then, is between (a) those views of inquiry which openly emphasise the 'separate' character of our 'abstract' knowledge, so conceiving 'theory' as a distinct, inevitably simplified and abstract account of a phenomenon; and (b) those views which conflate 'theory' and 'reality' so that a 'theory' (if given the name at all) is seen as a full and 'true' account of the reality in question, not an analytical simplification and selection. In effect, the one view focuses on the construction and/or testing of theories, while the other centres on providing a methodologically legitimated account of reality. So, commitment to (a) requires explicit formulation of our models and sentence systems, while (b) collapses construction, formulation, test, and explanation into the one interpretative process. Although it does not exhaust the range of possibilities, the extremes of this distinction might be usefully represented in the contrast between the received view's axiomatic calculus (a) and those conceptions which suggest that the world may be neutrally revealed by suitably qualified investigators (b).

There are overtones of several other common distinctions

here. In Reichenbach's terms, for instance, exclusive commitment to (a) suggests an undue emphasis on the 'context of justification', while (b) is similarly related to the 'context of discovery'. (4) Or, less philosophically sophisticated, the one approach involves an openly 'theoretical' inclination, while the other is more 'empirically' inclined, once a common distinction in sociological discourse, and one which has recently been extended and elaborated in the rather vague everyday use of terms like 'positivistic' and 'phenomenological'. Unfortunately, none of these terms, at least as they are usually formulated, suggests the full range of the contrast in question, and it is very difficult to find an appropriate pair. Some terminology is necessary, however, so I have returned to one of the oldest of procedural distinctions. I shall refer to emphasis on (a) as 'deductive', though it is important to recognise that this class of approaches is not exhausted by the axiomatised logical form of deductiv<u>ism</u>. It includes any perspective which centrally emphasises the separate analytic status of theory; the doctrinal form, deductiv<u>ism</u>, is only one such possibility. Its opposite, naturally enough, has to be 'inductive', a term which more aptly suggests the crucial characteristics of class (b). From an inductive point of view knowledge is not 'theoretical' in the sense that theory is a constitutive element of it open to subsequent empirical arbitration. Instead, we inductively infer our knowledge from observation and experience. Note, however, that this inductive/deductive distinction is not in practice absolute: a mixture of emphases is well within the bounds of possibility.

The second dimension of variation - designed to cross-cut

the inductive/deductive – is loosely derived from some of the views associated with Quine and Putnam. As we have seen, part of these philosophers' criticisms of empiricism rests on the recognition that not all theoretical statements relate directly to the empirical world. Accordingly, we can contrast those methodological views which claim knowledge to be demonstrable in atomised sentence terms with those in which what is at issue is a whole bundle of statements. Once more it is clear that mixed types are possible. Even Quine, who claims that knowledge relates to the external world mainly as a corporate body, does allow that some statements can indeed be subjected to individual tests. At the extreme, once more, is the received view. It can be understood as a claim that each individual proposition is, in principle, available for test, an emphasis which has led to the ritual of hypothesis testing playing a prominent part in social science claims to 'scientific' status. Finding appropriate terminology is less of a problem here: the opposition between <u>propositional</u> and <u>holistic</u> emphases will suffice.

Cross-classification, then, gives rise to a simple property space:

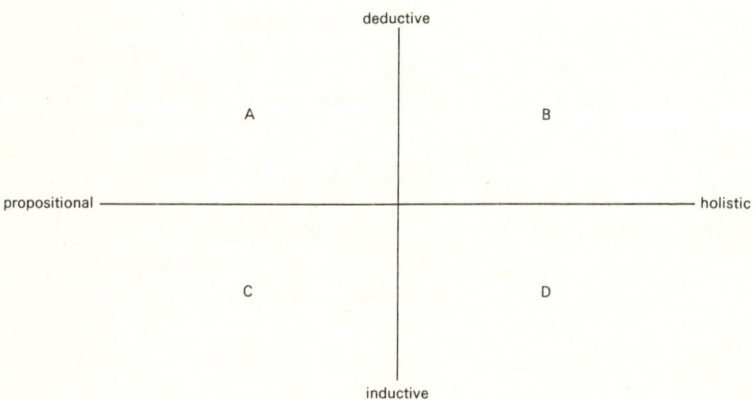

Figure 7.1 The axes of variation

Although a given methodology might be mapped anywhere within this property-space, many sociological perspectives actually crystallise around one or another of the quadrants. From my point of view, as I shall argue later, all four areas represent complementary elements of the process of inquiry, but, for the moment, it makes expository sense to consider each in turn, and to do so in relation to some familiar schools of sociological practice.

A Deductive-propositional: formal sociologies

Here central emphasis is on the abstract 'theoretical' character of scientific knowledge, combined with the condition that such knowledge be expressed and evaluated in propositional form. Pitched at the level of methodological doctrine, the paradigm case is that found in the most orthodox variants of the received view, where the axiomatic calculus is envisaged as the only legitimate form in which knowledge may be cast. In sociology this view has frequently figured as an implicit 'scientific' ideal, only occasionally surfacing as an open rationale. Some such case, for instance, is developed in Zetterberg's 'On Theory and Verification in Sociology', a work frequently cited by sociologists of a formal inclination. (5) At its simplest his argument is that the verbal formulations so common in sociology can be and should be axiomatised. Such reformulation, he suggests, serves to clarify the arguments involved and eliminate excess verbiage. So, for example, he restates Durkheim's famous account of the division of labour in terms of a set of axioms and a set of derived propositions.

Zetterberg's concern, then, is with propositional form in

general; others have added to that the claim that mathematics is the language in which such form is best expressed. But Zetterberg is not alone among sociological deductivists in stopping short of complete mathematicisation. As we saw in chapter 5, George Homans also advocates axiomatisation, a case succinctly expressed in his war-cry: 'no explanation without propositions!.' (6) For him the central sociological task is to offer explanations, for which deductive reasoning is deemed essential. Much sociological theory, he suggests, possesses 'every virtue except that of explaining anything', a jibe clearly directed at Parsons. Yet Homans is no less restricted. Like all exclusively formulated methodologies, Homans's version of 'science' illicitly elevates one feature of inquiry into holy writ, a strategy as crude as it is unpersuasive.

Fortunately, an extreme commitment to deductive-propositional form has not been common in actual sociological practice, though it is not unusual as an idealised element in methodological rhetoric. Nor has it been restricted to orthodox formalists in the Zetterberg tradition; there are also more recent perspectives which could easily lead to this style of inquiry. One could, for example, generate a formal sociology within the semiotic enterprise by viewing social life as a structured process of communication governed by <u>codes</u>. Codes, as Saussure and many later semiologists have indicated, are analysts' constructs, the application of which aids us in understanding seemingly diverse practices. If they were given formal expression (and there is an important strain in semiology toward such formalism) their use would loosely resemble that of 'laws' in orthodox deductivism: abstract covering propositions, separate from, though related to, the empirical domain. (7) The same could be said about some forms of structuralism and about those analyses of social action that emphasise the analytic task of establishing underlying 'rules' by which action is said to operate. So far, however, none of these more recently developed perspectives have formulated the deductive-propositional enterprise as straightforwardly as Zetterberg, Homans, or the various mathematical sociologies. Rigorous formalism is still rare in sociology.

B Deductive-holistic: abstract modelling

This class of perspectives shares with formal sociology a commitment to the abstract and analytic character of scientific knowledge, but without emphasising the centrality of propositional form. Instead, truth-claims relate to the analysis as a whole, hinging on its overall 'plausibility' or 'fit'. A great deal of so-called 'sociological theory' has fallen willy-nilly into this domain, though its most famous exemplar - Talcott Parsons - has been quite open in his espousal of deductive-holism. Throughout the development of the 'general theory of action' he argues in favour of a process of abstract theorising, a commitment underlined by his repeated references to Whitehead's 'fallacy of misplaced concreteness'. And although he has occasionally paid lip-service to the 'ideal' that theories should generate testable hypotheses, more often than not he appears to be satisfied with the claim that his models are a good 'fit' to the phenomena that they putatively model. Instead of deducing and testing discrete hypotheses, then, this style of sociology makes much more holistic claims, a practice to be found not only in Parsons's work but in almost all sociologies that operate at his level of abstract generality. (8)

Examples are many. Much sociological use of 'systems' terminology (frequently independent of Parsons's own version) is deductive-holistic, as have been other perspectives which emphasise the importance of emergent social structure. Althusserian Marxism, and, indeed, all those forms of structuralism which lean heavily on the <u>langue/parole</u> distinction, share a concern to develop theories that model the underlying processes of social life. In so doing they inevitably emphasise the abstract separateness of the theoretical enterprise, and they claim that a theory must be taken as a whole. Its component elements cannot be considered in isolation from the overall theoretical structure.

C Inductive-propositional: systematic empiricism

If abstract modelling has been a constant feature of the sociological landscape, 'systematic empiricism' (a term borrowed from the Willers) has been its very bedrock. (9) It's easy to see the attraction. Simple propositional form, with little or no emphasis on precise logical interrelations, keeps sociology superficially 'close' to the empirical world (in contrast to the abstract edifices of the 'modellers') and makes some gesture toward the orthodox ideal of testing individual hypotheses. But such gestures are superficial. That 'ideal', at least as it has been promulgated in sociology, is something of a vulgarisation of a perspective more concisely realised in the work of formal sociologists. The hypotheses of systematic empiricism are not actually subjected to tests: rather, they are empirical generalisations inductively inferred from observation and then ritually recast as 'hypotheses'. It is this superficial allegiance to 'scientific method' that leads the Willers to dismiss such enterprises as 'pseudo-science', and though the expression is apt the dismissal is perhaps a little harsh. Systematically developing a set of empirical generalisations is an essential element in empirical inquiry: without some such substance sociology would indeed be the arid enterprise commonly detected by critics of Parsonian theory. Systematic empiricism only becomes pseudo-science at the point at which it purports to have exhausted legitimate investigation. As part of the process of generating 'described phenomena' such work is essential - but it is only a beginning, not a conclusion.

Far and away the crudest and most revealing tool of the sociological systematic empiricist has been the sample survey, avidly dredging for data, and often trying to render that data respectable by post-hoc reformulation in hypothesis terms. By cross-tabulating each item with every other one such methodologies create a set of empirical generalisations legitimated in a patina of statistical and 'scientific' sophistication. But what is at issue is much the same in all methods of developing empirical generalisations, be they in historical research, exploratory experimentation, Mertonian functionalism, participant-observation studies, or even in the apparently more formally inclined work of ethnomethodologists and conversation analysts. Though these various approaches differ radically in some respects, they resemble each other in their concern to generate empirical grist for the analytic mill. If they are exclusively inductive-propositional (and that is true of some), then they make the mistake of limiting the whole research process to the provision of grist. By applying his or her preferred methods, be they statistical, functionalist, 'phenomenological', or whatever, the systematic empiricist elevates tentative claims into truth, research findings into established knowledge. (10) 'Research has shown', that favourite catch-phrase of academic journalism, conceals a multitude of methodological sins.

D Inductive-holistic: interpretative sociologies

Like all inductive perspectives inductive-holism claims to grasp the world directly, but at a rather more ambitious level than that of systematic empiricism. Instead of dealing in individual generalisations, whole bundles are gathered together into total interpretations. Interpretative sociologies propose a comprehensive account, and they seek to evaluate it in terms of its overall interpretative capacity and not the alleged 'accuracy' of its statements taken individually. Once 'established', furthermore, this interpretative account stands without further arbitration, for, like all such holistic social philosophies, it is its general plausibility which is at issue, not its demonstrable precision and use in relation to some externally given reality. In a word, it is a 'world-view'.

A pure example of such a sociology is almost impossible to find. That is not surprising. If any enterprise has been constantly castigated in twentieth-century sociology it is that of advancing an 'unscientific' world-view. Sociologists, who have agreed on nothing else, coincide in dismissing such exercises to the value-laden domain of social philosophy. Yet there are strong elements of inductive-holism in a number of sociological perspectives. Berger's work, for example, combines a general interpretative account (a 'philosophical anthropology') with specific empirical claims. Whether his concern is knowledge, religion, or modernisation, he asks us to accept his total view as plausible, bolstering that plausibility by reference to selected empirical generalisations. (11)

Nor is this strategy only to be found in work as macroscopic as Berger's. Sociologists of very different persuasions - Goffman, for instance - make similar demands. In analysing the minutiae of face-to-face interaction Goffman, too, offers us both an holistic interpretation of social life and a selection of empirical observations. Its plausibility is a function of how we, the readers, respond to his work, how we relate it to our experience, a 'literary' feature which has caused more orthodox sociologists to question the very sociological status of Goffman's enterprise. (12) This emphasis on the individual reader's role in judging the 'validity' of an interpretation is common to all inductive-holistic approaches. Since the interpretation has been inferred from the analyst's perceptions ('methodologised' or not) we can only say whether this account gells with our own experience. All sociologies in this quadrant thus place great methodological emphasis on the role of 'experience' in sociological inquiry, neglecting, presupposing, or even denying the possibility of a separate 'test' of their claims. In so doing they collapse the various features of inquiry into one interpretative process.

COMPLEMENTARITY AND CONFLICT

As the very scantiness and superficiality of my examples should suggest, the point of this classification is not to throw light on individual sociologies (for which purpose there are much better approaches) but to illuminate the process of sociological inquiry in general. I shall try to do so by relating the terms of this analysis of methodological variation to some of the features of inquiry as they have been outlined elsewhere in this book. In so doing I want to suggest (if not actually establish) that all four 'methodologies' only function properly in combination. Each presumes the sort of 'knowledge' developed by others. Accordingly, we need to combine the seeming alternatives of formal sociology, abstract modelling, systematic empiricism, and interpretative sociology into a working understanding of the total enterprise. From this point of view, then, it has been unfortunate for sociology that so many of its practitioners have been devoted to establishing exclusive rights over their own methodological domains, defending these limited boundaries with short-sighted vigour. Such exclusivism has surely proved counter-productive.

We can begin to see what might be involved in a synthesis of methodologies by exploring the ways in which the different sociologies have over-emphasised particular features of the process of inquiry. Because it is so widespread, first consider systematic empiricism. Here, as we have seen, the major focus is on a largely unstructured, ad hoc sentence system. It is the substantive content of propositions that is important; not their structured interrelation. The statements of systematic empiricism are deemed to be methodologically warranted representations of the real world, and their apparent 'obviousness' has often prompted the charge that sociology deals only in 'trivial' generalisations. To transcend the limitations of this apparent concern with isolated 'facts', ad hoc sociological propositions require mapping into a model. The simplest versions of this operation usually invoke the metaphysics of causation; connectives in the propositions are translated directly into causal relations. At its extreme this form of reasoning imputes cause on the basis of specified statistical associations. Blalock's work is classic here, and there is something to be said for the claim that much notionally causal analysis in sociology, Blalock included, is the last resort of a desperate empiricism. (13) By tacitly assuming a particular account of 'cause' such approaches can avoid any overt reference to theory and so let the data apparently 'speak for itself'. But to impute cause is to make a claim about the nature of the association expressed in an empirical generalisation, and to do that is, explicitly or implicitly, to invoke a model of the process in question.

A similar misconception arises in a slightly different form in that other traditional stronghold of sociological systematic empiricism: Mertonian functionalism. (14) Here, too, the analyst looks at the world and produces a roughly propositional account of it. But this time the external reference is not to <u>cause</u> but to <u>function</u>: the aim is to demonstrate the functionality of relations in the real world. Evidence is selected against this criterion, usually without any reference to the fact that the cogency of such analyses derives implicitly from a more general model of system functioning.

For the concept _function_, whatever its other failings, has no meaning unless we first identify system needs and build them into a coherent model. If we do not, then Mertonian functionalism reduces to the most pernicious of ad hoccery: find a problematic item and invent a need for which it is functional. If necessary, of course, one can go on inventing needs just as long as there are problematic items. Without an _explicit_ model, analyses such as this answer no questions and solve no problems. Thus, although systematic empiricism, of whatever form, might provide us with a body of 'described phenomena', if we are to make any further analytic progress we need to relate such materials to an appropriate model.

Similar considerations apply to formal sociology in as much as it, too, over-states the importance of the sentence system. Here, however, claims are pitched at a more abstract level. Where systematic empiricism deals mainly in loosely interrelated low-level empirical generalisations, formal sociologies operate with propositions of varying degrees of abstraction interrelated rather more precisely. Indeed, it is the formal structure of this interrelation which commands most attention, and straightforward generalisations from data play a subsidiary role. The formal sociologist's task is to construct new theories, or translate old ones, such that they fit the ideal deductive form. On this account, of course, all theories, if they are to be properly so called, must be deductive, for it is only through deductive 'covering' that we are able to offer precise 'scientific' explanations. As I have already argued, this is a fatally limiting view, though it is true that the relation of propositional 'covering' is an important element in explanation. Sentence systems are not only as they appear in systematic empiricism: assemblies of empirical generalisations. They can also incorporate more abstract propositions, and where their terms do not derive meaning solely or directly from observational categories, they relate back, network fashion, to an underlying model. It is in this way that a model can potentially 'cover' the lowest of empirical generalisations.

In sociology, however, models have usually been developed independent of concrete explanatory exigencies. There is nothing necessarily wrong with that. Strategies for model building are many and varied, and problems arise not about the construction of a model but about its status as a finished product. However elegant or precise it may be, a model is castrated if it is never related to the empirical domain which it putatively models. Far too much abstract modelling in sociology has failed properly to establish the complex and partly propositional relations that link models to 'described phenomena'. This is not to claim, however, that sociological theories are too abstract, too 'far from reality'; in many ways that is their virtue. No one should sensibly expect that all the terms and statements of a model bear directly on the phenomenal world. But it is important to clarify the points of linkage, the loci at which available sentence systems may be fed into a model. Otherwise any sense of empirical reference remains dangerously undefined.

I have mentioned Parsons as a paradigm of abstract modelling in sociology. One of the more common criticisms of his work has been to claim that his elaborate theorising is empirically empty, and, in as much as he fails to clarify the points at which we might reasonably expect his model to relate to relevant sentence systems, then the charge is plausible enough. We are left with the feeling that his work is of no empirical relevance, a worry which is in no way assuaged when we come to consider Parsons's own attempts at application. Whatever their plausibility (and some of them are plausible) his more 'concrete' claims remain independent of his models to the degree to which precise connections are unclear. One possible solution, I suppose, would be an alliance of Parsonian modelling with a suitably delimited formal sociology. It is a measure of the discipline's intellectual defensiveness that such a proposal would be greeted as abject miscegenation, most especially by Parsonians and formal sociologists.

Parsons, of course, is not merely an abstract modeller: a great deal of his work can be seen as an attempt to elaborate a language and conceptual scheme for sociological theory. He shares that concern with the various more interpretatively inclined sociologies, though they differ in asserting a direct link between observed phenomena and interpretative account. For them it is sufficient to develop a language and a conceptual scheme which, directly applied to the social world, provide appropriate interpretations. It is taken for granted that interpretative application is unproblematic; that there is no need for us separately to formulate models and sentence systems. It is this short-circuiting of the process of inquiry that lends interpretative sociologies their air of legitimate generalisation: only when armed with the right conceptual apparatus, they imply, is one in a position to render reality intelligible. But, as we have seen, a conceptual scheme can be specified in more than one model, some of which will be inconsistent with others. Similarly, a model may in its turn 'cover' a whole series of sentence systems, with the resulting complex of relationships represented schematically in Figure 4.2. Any account which is offered holistically as a direct sociological interpretation, therefore, is keeping covert these crucial elements in the process of analysis. And if they are covert they are hardly open to critical discussion, a difficulty which has meant that much argument in 'sociological theory' has had to revert to argument about very general basic assumptions.

This question of interpretation, though has two levels. In the one, difficulties arise from the ways in which we (as sociologists) process our 'experience' so as to generate conceptual schemes which then, inevitably, modify that experience. At the second level, different, though not unrelated, difficulties arise from the more specific task of producing what we are prepared to accept as 'described phenomena' or 'data'. It is by now almost a cliche to suggest that our descriptions are theory-laden, that we employ what Lakatos calls 'interpretative' or 'auxiliary' theories to generate our data. But on what grounds do we choose among such theories? What criteria do we employ in deciding which framework we will use to process the phenomenal world? Lakatos suggests that 'the methodological falsificationist uses our most successful theories as extensions of our senses.' (15) If this is the case - and it would be difficult to deny - the basic question still remains: how can we decide, whether we are falsificationists or not, which of our theories is sufficiently 'successful' to merit such treatment. In chapter 6 I have already suggested that the criteria of choice advanced in the Popperian tradition are not adequate to the range of scientific practice. We take decisions about the adoption of particular interpretative theories on a variety of grounds, formal and informal, and including reference to our values, interests, and commitments as well as more seemingly 'rational' criteria.

Furthermore, the problem is additionally complicated by the fact that in sociology our 'data' is often already processed in terms of frameworks taken-for-granted in everyday accounts. Ethnomethodology has expressed one of its fundamental commitments in the claim that we should explore as a _topic_ what conventional sociology considers to be a _resource_. (16) That is, the accounts offered by participants in social processes, frequently taken as unproblematic sources of data by sociological researchers, should themselves be systematically studied. Such analysis, in the terms employed in this discussion, would constitute an inquiry into the interpretative theories that social actors utilise in everyday understanding. Ethnomethodology has been very influential in making a positive attempt to grapple with such difficulties of interpretation, and the ethnomethodological critique of the constitution of sociological data is a cogent one. But this should not be allowed to disguise the fact that ethnomethodologists, too, utilise interpretative theories to constitute _their_ data, theories which presuppose a certain image of human action, its reflexivity, and whatever part is played in it by such features as reasons, intentions, and motives. There is no way in which our empirical materials can be somehow drained of their theoretical taint; there is no absolute empirical foundation. To borrow Neurath's metaphor, the sociological enterprise – like all forms of empirical inquiry – requires us to build the boat in which we are sailing while we are actually sailing it. Recently renewed interest in hermeneutics, in the problems of verstehen, and in general difficulties of interpreting social action, re-emphasise one feature of sociological inquiry which had suffered neglect in post-war research. But that interest, and the work that it has occasioned, does not by itself constitute an adequate sociology. It is only one piece in the jigsaw of complementarity.

SOCIOLOGY, NATURALISM, AND THE PHILOSOPHY OF SCIENCE

Throughout this discussion I have presumed that we already have at our disposal the elements of a 'scientific' sociology, but separated one from another in partial perspectives. This is not simply woolly pluralism on my part. In developing this chapter's simple typology of sociological methodologies I have sought to show how sociologists have reified particular elements of inquiry. It is in this general sense, then, that I am a 'naturalist'. I believe that we can meaningfully speak of a 'structure' of inquiry, that the philosophy of science is an essential locus for the analysis of that structure, and that, in as much as it is 'scientific' to work within such a structure, there can be a 'scientific sociology'. Nothing much hangs on that, however, since there is surely no longer any need to justify sociological work by seeking to demonstrate that it is scientific. My 'naturalism' is not part of a philosophy-of-science justification. Rather, I believe that in utilising some of the terms of philosophy of science we can better understand the sorts of claims which can and can't be made for sociological work, better grasp the interrelations between its various methodological practices.

There is, of course, a well-established tradition which would deny this and any other naturalist claim, however limited. From this point of view a radical distinction must be made between natural and human sciences based on the distinctive features of human action. The argument takes many forms, but common to most of them is the view that natural science is practised more or less as orthodox philosophy of science suggests, and that, because social action is such that we can only ever hope to interpret it, scientific methods are inapplicable. An obvious response to this view is to argue that orthodox philosophy of science is inaccurate, and that more recent work develops an account of science which can be legitimately applied to sociology. Another not unrelated alternative is to argue that antinaturalists are not merely mistaken about science; they also misconceive sociology's subject matter, which is not such as to demand non-scientific forms of analysis. One can accept, that is, that the social actor is an active, meaning-endowing, reflexive subject, without also claiming that 'a naturalist methodology based on the student as observer has to "objectify" the actor under study in some pernicious epistemological-cum-moral sense.' (17) A radical variation of this thesis has been developed by Bhaskar, and it is instructive to consider the general form of his argument. (18)

Chapter 7

Bhaskar's case is for a 'qualified, anti-positivistic naturalism' which none the less retains the view common to anti-naturalists that 'it is the nature of the object that determines the form of its science.' He thus seeks to define sociology's object of study, and show that, thus defined, that object is amenable to scientific analysis. In so doing he does not claim that the distinctive character of sociology's object is of no methodological significance; it does place limits on naturalism and these limits can be defined. The pillars of his argument, then, are two-fold: a 'realist' account of science in which the central goal of scientific inquiry is to move from manifest phenomena to an understanding of the structures from which they are generated; and a distinctive view of sociology's object of study. This, he says, has to be understood in both <u>relational</u> and <u>transformational</u> terms. That is to say, sociology is concerned 'with the persistent <u>relations</u> between individuals (and groups), and with the relation between those relations' in the context of a view of social actors as reproducing and transforming these persistent structures. (19)

> men in their social activity must perform a double function: they must not only make social products but make the conditions of their making, i.e. reproduce (or to a greater or lesser extent transform) the structures governing their substantive activities of production. Because social structures are themselves social products, they are themselves possible objects of transformation and so may be only relatively enduring. And because social activities are interdependent, social structures may be only relatively autonomous. Society may thus be con-

ceived as an articulated ensemble of such relatively independent and enduring structures; that is, as a complex totality subject to change both in its components and their interrelations. (20)

For Bhaskar, then, there are emergent social properties which allow us to define sociology's object 'society' in this way. Thus conceived, of course, 'society' is a theoretical object in the sense that it cannot be encountered other than through its effects. But that, as he observes, does not run counter to naturalism. The limits on naturalism that flow from this conception relate rather to the fact that social process is 'open': social structure cannot be understood in terms of invariant regularities since it is definitionally open to transformation. And this, in turn, means that there can be no decisive empirical tests, and that sociological theory is inevitably incomplete. It is the explanatory force of our theories which should determine our selection among them, not their predictive capacity, and in constructing such theories we are inevitably involved in seeking to transcend the explanations already found within society.

I have no wish to form a judgment on Bhaskar's case, and anyway I could hardly do so on the basis of a summary as cryptic as this. I have outlined it here only to demonstrate the possibility of alternatives to the conventional opposition between 'positivist' naturalism and hermeneutic or interpretative exclusivism. Whether in terms of 'interpretive' versus 'normative', 'order' versus 'control', or 'positivistic' versus 'phenomenological', it has become common in sociology to claim that the fact that social actors are active agents necessitates a non-naturalistic social study. (21)

One of the disturbing features of such claims is that they invariably lead to a neglect of <u>structure</u> in favour of a subjectivist viewpoint. Bhaskar seeks a conception of sociology which will allow us to conceive the social actor as active ('transformational') and which will combine that recognition with an analysis of emergent structure. In so doing he seeks to transcend the conventional oppositions of the naturalism debate, thus re-engaging the discourse of the philosophy of science with that of sociological theory.

Nor is he alone in this attempt to rescue sociology's concept of social structure from both positivist reification and interpretative reduction. Giddens, too, has attempted conceptually to overcome the opposition between what he refers to as 'voluntarism' and 'determinism' and to develop a 'theory of structuration'. (22) This task seems to me to be central to the further development of sociological inquiry. Science, as I have tried to suggest throughout this study, is not the uni-dimensional pursuit it has been proclaimed to be by those seeking methodological saviours. It involves a range of practices suspended in a network of rules and theories, and, in my conception of it, it can easily encompass the methodological plurality of contemporary sociology. Unfortunately, sociology's past insecurity has meant that all too many sociological practices have vied with each other for the title of proper social science, an inclination which has reached disturbing proportions in the fragmentation of the 1970s. The key to reuniting methodologically the sociological enterprise has, I believe, two principal features. We need to take seriously the claim that the variety of sociological methods reflect potentially complementary activities rather than fundamentally irresolvable differences in perspective. And, in taking that claim seriously, we must recognise that any basis for unification requires an understanding of sociology's subject matter in terms of both social structure and active agency. It's not as if this is merely a matter of abstruse theoretical interest. The very data that we employ, the 'described phenomena' that are so central to any sense of empirical reference, depend upon our interpretative theories. If those theories continue to reify either structure or action, then our data will continue to reflect that reification, and sociology will remain fragmented in the most fundamental sense.

Philosophy of science can help us here. It cannot do the job for us; it isn't the philosopher's stone so long sought by sociologists. But it can help us to clarify quite what is involved in empirical inquiry, and what limits must be imposed upon the claims that we make. In offering us a kind of map, a framework within which to conceptualise our activities, it can guide us, though no more than that, to the places at which we must forge new connections. It can help us to see that many of the arguments traditionally marshalled in opposition to science in sociology do not confound systematic sociology itself, but rather demand a new synthesis and a wider reflexive understanding. Needless to say this book has had no such grandiose pretensions; for me it will be sufficient if I have persuaded anyone that the subject is worth further consideration.

Notes

CHAPTER 1 THE PHILOSOPHER'S STONE

1. There are many standard texts, even the best displaying some of these problems. As far as a distinction can clearly be drawn, for those produced by philosophers see, as examples: Alan Ryan, 'The Philosophy of the Social Sciences', Macmillan, London, 1970; Richard Rudner, 'Philosophy of Social Science', Prentice-Hall, Englewood Cliffs, 1966; and, perhaps the best currently available, Richard J. Bernstein, 'The Restructuring of Social and Political Theory', Blackwell, Oxford, 1976 and Methuen, London, 1979. Inevitably Peter Winch, 'The Idea of a Social Science', Routledge & Kegan Paul, London, 1958, must be mentioned here though it hardly sets out to be a 'text'. However, it is arguably a case of 'philosophical imperialism'. As examples from sociologists see: Michael Lessnoff, 'The Structure of Social Science', George Allen & Unwin, London, 1974; Walter Wallace, 'The Logic of Science in Sociology', Aldine-Atherton, Chicago and New York, 1971; David Willer, 'Scientific Sociology', Prentice-Hall, Englewood Cliffs, 1967. Recently sociologists' discussions have ranged rather more widely. See, for example, Ted Benton, 'Philosophical Foundations of the Three Sociologies', Routledge & Kegan Paul, London, Henley, Boston, 1977; Russell Keat and John Urry, 'Social Theory as Science', Routledge & Kegan Paul, London and Boston, 1975; Barry Hindess, 'Philosophy and Methodology in the Social Sciences', Harvester Press, Brighton, 1977.
2. As we shall see this shift has had widespread implications in the philosophy of science. Meanwhile, for a

stimulating discussion of the general question of epistemology (though perhaps finally too psychologically disposed for a sociologist's taste) see W.V. Quine, Epistemology Naturalised, chapter 3 of his 'Ontological Relativity and Other Essays', Columbia University Press, New York and London, 1969.

3 For a very helpful discussion of the metaphysics and general character of positivism see R. Harré, 'Theories and Things', Sheed & Ward, London, 1961. On p. 10 he remarks that: 'the common form of all positivist doctrines is that they recommend that for the purpose of science we adopt only one ontological class.' For some more general uses of the term see the introduction and materials collected in Anthony Giddens (ed.), 'Positivism and Sociology', Heinemann, London, 1974. For an excellent general discussion see Anthony Giddens, Positivism and its Critics, in Tom Bottomore and Robert Nisbet (eds), 'A History of Sociological Analysis', Heinemann, London, 1979.

4 The suggestion that sociology should be 'demystificatory' has been common currency for some time. It was thoroughly revived by John Rex in his 'Sociology and the Demystification of the Modern World', Routledge & Kegan Paul, London and Boston, 1974.

5 Sidney Morganbesser, Is it a Science?, 'Social Research', 33, 2, 1966, reprinted in Dorothy Emmet and Alasdair MacIntyre (eds), 'Sociological Theory and Philosophical Analysis', Macmillan, London, 1970.

6 The sociology of science has now accumulated a great deal of evidence for this general claim, although it is the impressionistic account of Thomas Kuhn's 'Structure of Scientific Revolutions', Chicago University Press, Chicago, 1962, which has made the most 'public' impact.

7 Alison Lurie, 'Imaginary Friends', Pan, London, 1968. She also refers to 'boxes and arrows' but that is perhaps closer to home!

8 David Willer and Judith Willer, 'Systematic Empiricism: Critique of a Pseudo-Science', Prentice-Hall, Englewood Cliffs, 1973. Also see David Willer, op. cit., and Judith Willer, 'The Social Determination of Knowledge', Prentice-Hall, Englewood Cliffs, 1971.

9 For the term 'modest empiricism' see Israel Scheffler, Prospects of a Modest Empiricism, 'The Review of Metaphysics', X, 3-4, 1957.

CHAPTER 2 PRACTICAL EPISTEMOLOGY

1. For Popper's classic statements see Karl Popper, 'The Logic of Scientific Discovery', Hutchinson, London, 1959, and 'Conjectures and Refutations', Routledge & Kegan Paul, London, 1963. His more recent views are represented in a collection of his essays: 'Objective Knowledge', Oxford University Press, 1972. He has been very influential and his views have been widely diffused and more or less subtly changed. John Watkins is possibly the most consistently 'pure' Popperian, contributing too much over too many years to be listed here. There seems no question, however, that modern philosophy of science is shifting away from the Popperian mode. For some sense of this shift see the two volumes derived from the 1975 Kronberg conference: Gerard Radnitzky and Gunnar Andersson (eds), 'Progress and Rationality in Science' and 'The Structure and Development of Science', Reidel, Dordrecht, Boston, London, 1978 and 1979. See also the lucid discussion in Stefan Amsterdamski, 'Between Experience and Metaphysics', Reidel, Dordrecht, Boston, 1975. I shall focus on the Popperian tradition in chapter 6.
2. As a good example of 'rational reconstruction' take the work of Carl Hempel, perhaps most accessibly discovered in his collection of essays: 'Aspects of Scientific Explanation', Free Press, New York, 1965. See also his elementary introduction: 'Philosophy of Natural Science', Prentice-Hall, Englewood Cliffs, 1968. Another well-known figure is Ernest Nagel. See his classic text: 'The Structure of Science', Routledge & Kegan Paul, London, 1961, and his essays collected in 'Logic Without Metaphysics', Free Press, Chicago, 1956.
3. Nelson Goodman, 'Fact, Fiction and Forecast' (second edition), Bobbs-Merrill, New York, 1965, p. 47.
4. In the philosophy of science see Thomas Kuhn, 'The Structure of Scientific Revolutions', University of Chicago Press, 1962, second edition, 1971. See also his contributions to a critical symposium on his work: Imre Lakatos and Alan Musgrave (eds), 'Criticism and the Growth of Knowledge', Cambridge University Press, London, 1970. Feyerabend has written widely, but his views can perhaps be best represented by the following: Paul K. Feyerabend, Against Method: Outline of an Anarchistic Theory of Knowledge, 'Minnesota Studies in

the Philosophy of Science', Minnesota University Press, Minneapolis, 1970; Consolations for the Specialist, in Imre Lakatos and Alan Musgrave (eds), op. cit; Problems of Empiricism, in Robert G. Colodny (ed.), 'Beyond the Edge of Certainty', Prentice-Hall, Englewood Cliffs, 1965; Problems of Empiricism, Part II, in Robert G. Colodny (ed.), 'The Nature and Function of Scientific Theories', University of Pittsburgh Press, 1970; 'Against Method', New Left Books, London, 1975; On the Critique of Scientific Reason, in R.S. Cohen, P.K. Feyerabend and M.W. Wartofsky (eds), 'Essays in Memory of Imre Lakatos', Reidel, Dordrecht, Boston, 1976.

5 This refers to the debate over the 'covering law' approach to explanation advanced by Carl Hempel. As I shall argue in chapter 5 his position changes quite considerably such as to limit the strictly _logical_ characteristics of his analysis.

6 For a classic discussion see W.V. Quine, Two Dogmas of Empiricism, most accessibly available in his 'From a Logical Point of View', Harper & Row, New York, 1963.

7 I am not sure that Ryle would approve of this version of his famous distinction. See Gilbert Ryle, 'The Concept of Mind', London, 1949.

8 One such appeal, much discussed, is to be found in Alvin W. Gouldner, 'The Coming Crisis of Western Sociology', Basic Books, New York and London, 1970. However, his specific discussion of this topic, limited to one chapter at the end of the book, does not probe very far. For a more extensive consideration see Roland Robertson, The Sociocultural Significance of Sociology: A Reconnaissance, in A.H. Hanson, T. Nossiter and Stein Rokkan (eds), 'Imagination and Precision in Political Analysis', Faber & Faber, London, 1973.

9 For an excellent general introduction see D.W. Hamlyn, 'The Theory of Knowledge', Macmillan, London, 1970. For a more complex discussion in a somewhat different mould see Arthur C. Danto, 'Analytical Philosophy of Knowledge', Cambridge University Press, London, 1968. For an introduction much concerned with the rise of pragmatism see Bruce Aune, 'Rationalism, Empiricism, and Pragmatism: An Introduction', Random House, New York, 1970. See also the much quoted A.J. Ayer, 'The Problem of Knowledge', Penguin, Harmondsworth, 1956, and his excellent introductory philosophy text, 'The Cen-

tral Questions of Philosophy', Weidenfeld & Nicolson, London, 1973.
10 W.V. Quine, 'Word and Object', MIT Press, Cambridge (Mass.), 1960, p. 3.
11 For useful discussions of the relation between belief and knowledge see D.W. Hamlyn, op. cit., chapter 4, and Arthur C. Danto, op. cit., chapter 4. See also the essays collected in A. Phillips Griffiths (ed.), 'Knowledge and Belief', Oxford University Press, London, 1967.
12 See Karl Popper, 'Objective Knowledge', op. cit., especially, in this context, his unhappiness with 'the commonsense theory of knowledge' evinced in his essay Two Faces of Common Sense: an Argument for Commonsense Realism and Against the Commonsense Theory of Knowledge. What Popper identifies as the commonsense theory is a long way from my usage of that term. See also the arguments advanced in Arthur C. Danto, op. cit.
13 Alison Lurie, 'Imaginary Friends', Pan Books, London, 1970.
14 Holzner speaks of an epistemic community in the following terms: 'All members of such a community, in their capacity as members agree on "the" proper perspective for the construction of reality. In these communities the conditions of reliability and validity of reality constructs are known and the applicable standards are shared.' Burkart Holzner, 'Reality Construction in Society', Schenkman, Cambridge (Mass.), revised edition, 1972, p. 69.
15 D.W. Hamlyn, op. cit., p. 118.
16 The phrase is borrowed from Tom Stoppard's philosophically sceptical and hilarious play, 'Rosencrantz and Guildenstern are Dead', Faber & Faber, 1967, p. 79.
17 Thomas Kuhn, 'The Structure of Scientific Revolutions', op. cit.
18 Thomas Kuhn, Reflections on My Critics, in Imre Lakatos and Alan Musgrave (eds), op. cit., p. 231. See also John Watkins, Against Normal Science, in the same volume. This collection is a useful reflection of philosophy of science responses to Kuhn's works.
19 See, for example, Imre Lakatos, Falsification and the Methodology of Scientific Research Programmes, in Imre Lakatos and Alan Musgrave (eds), op. cit.
20 Paul K. Feyerabend, Against Method: Outline of an

Anarchistic Theory of Knowledge, op. cit. Also the two Problems of Empiricism papers, op. cit. Much of this section was written before the appearance of Feyerabend's book-length assembly of his views in 'Against Method', op. cit. I have here retained the page references to the original Against Method article since the book is not substantially different - at least, not on the topics discussed here.

21 Paul K. Feyerabend, Against Method, op. cit., p. 21.
22 Derek L. Phillips, 'Abandoning Method', Jossey-Bass, London, 1973, chapters 7, 8, and 9.
23 See, for example, Feyerabend, On the Critique of Scientific Reason, op. cit.; his In defence of Aristotle: Comments on the Condition of Content Increase, in Gerard Radnitzky and Gunnar Andersson (eds), 'Progress and Rationality in Science', op. cit.; also see the polemical rejoinder The Gong Show - Popperian Style in the same volume.
24 The term is adapted from Peter Berger's text on the sociology of religion. The American edition of his 'The Social Reality of Religion', Faber & Faber, London, 1969, is more aptly titled: 'The Sacred Canopy'.
25 This is a major point of many of Feyerabend's examples in Against Method, op. cit., and Problems of Empiricism, part II, op. cit.
26 Classic Baconian inductivism is well known, but for a brief and illuminating account of the subject see P.B. Medawar, 'Induction and Intuition in Scientific Thought', Methuen, London, 1969. On Popper see, especially, his 'Logic of Scientific Discovery' and 'Conjectures and Refutations', op. cit. For Hempel's early 'over-deductivist' view see Carl J. Hempel and Paul Oppenheim, Studies in the Logic of Explanation, 'Philosophy of Science', 15, 1948, reprinted in his 'Aspects of Scientific Explanation', op. cit.
27 The phrase is here borrowed from Quine and Ullian's engaging elementary introduction: W.V. Quine and J.S. Ullian, 'The Web of Belief', Random House, New York, 1970, though its origins, of course, lie in anthropology. See, notably, E.E. Evans-Pritchard, 'Witchcraft, Oracles and Magic among the Azande', Oxford University Press, 1937.
28 W.V. Quine, 'Word and Object', op. cit., pp. 24-5.
29 W.V. Quine, Two Dogmas of Empiricism, op. cit., pp. 42-3.

CHAPTER 3 THEORIES AND THINGS

1. I have here borrowed the title of Rom Harré's excellent study. See R. Harré, 'Theories and Things', Sheed & Ward, London, New York, 1961.
2. For examples of the increasing range of dispute (as opposed to argument within agreed perspectives) see the following collections: Robert G. Colodny (ed.), 'The Nature and Function of Scientific Theories', University of Pittsburgh Press, 1970; Imre Lakatos and Alan Musgrave (eds), 'Criticism and the Growth of Knowledge', Cambridge University Press, 1970; Michael Radner and Stephen Winokur (eds), 'Minnesota Studies in the Philosophy of Science', vol. IV, University of Minnesota Press, Minneapolis, 1970; Frederick Suppe (ed.), 'The Structure of Scientific Theories', University of Illinois Press, Urbana, Chicago, London, 1974; Robert S. Cohen and Marx W. Wartofsky (eds), 'Methodological and Historical Essays in the Natural and Social Sciences', Reidel, Dordrecht, Boston, 1974; Gerard Radnitzky and Gunnar Andersson (eds), 'Progress and Rationality in Science', Reidel, Dordrecht, Boston, London, 1978. Compare these volumes with any of the older traditional collections, e.g. the volume of introductory readings: Arthur Danto and Sidney Morgenbesser (eds), 'Philosophy of Science', World Publishing Co., Cleveland and New York, 1960. For a useful general account of developments in modern philosophy of science see Stefan Amsterdamski, 'Between Experience and Metaphysics', Reidel, Dordrecht, Boston, 1975.
3. Stephen Toulmin, The Structure of Scientific Theories, in Frederick Suppe, op. cit., p. 613.
4. Ibid., pp. 612-13.
5. Hilary Putnam, What Theories are Not, in Ernest Nagel, Patrick Suppes, and Alfred Tarski (eds), 'Logic, Methodology and Philosophy of Science, Stanford University Press, 1962.
6. Frederick Suppe, The Search for Philosophic Understanding of Scientific Theories, in Frederick Suppe, op. cit.
7. See, for example, the changes of position and responses to criticism reflected in his collection of essays: Carl G. Hempel, 'Aspects of Scientific Explanation', Free Press, New York, 1965. See also the discussion in Carl G. Hempel, On the 'Standard Conception' of Scien-

tific Theories, in Michael Radner and Stephen Winokur (eds), op. cit.
8 See Carl G. Hempel, The Theoretician's Dilemma: A Study in the logic of Theory Construction, in Carl G. Hempel, op. cit. It is notable that the late 1950s in British sociology saw very little serious consideration of 'theory' in terms other than this. John Rex's 'Key Problems of Sociological Theory', Routledge & Kegan Paul, London, 1961, was the only serious British-produced discussion of contemporary sociological theory.
9 See Frederick Suppe, op. cit., p. 16 for a summary statement; compare this with his 'final version' of the received view on pp. 50-2. For a useful general introductory discussion see R. Harré, 'The Philosophies of Science', Oxford University Press, London, Oxford, New York, 1972. For other general discussions of the received view see: Hilary Putnam, op. cit.; Peter Achinstein, 'Concepts of Science', John Hopkins Press, Baltimore, 1968; Herbert Feigl, The 'Orthodox' View of Theories: Remarks in Defence as well as Critique, in Michael Radner and Stephen Winokur (eds), op. cit.; Carl G. Hempel, On the 'Standard Conception' of Scientific Theories, ibid., as well as many other recent papers and volumes.
10 Hilary Putnam, op. cit., p. 240.
11 Ibid.
12 Carl G. Hempel, On the 'Standard Conception' of Scientific Theories, op. cit., p. 146.
13 See Hans L. Zetterberg, 'On Theory and Verification in Sociology', Bedminster Press, Totlowa, NJ, 1965; see also George C. Homans, 'Social Behaviour: Its Elementary Forms', Routledge & Kegan Paul, London, 1961, and his 'The Nature of Social Science', Harcourt Brace Jovanovich, New York, 1967.
14 Frederick Suppe, op. cit., p. 52.
15 Ibid., pp. 62-6. He discusses the question primarily in relation to the axiomatisation and formalisation issues.
16 The term is adapted from Northrop's 'epistemic correlations', though I do not intend it to carry the received view overtones it carries in his usage. See F.S.C. Northrop, 'The Logic of the Sciences and the Humanities', Macmillan, New York, 1947.
17 Carl G. Hempel, On the 'Standard Conception' of Scientific Theories, op. cit., p. 148.
18 Ernest Nagel, 'The Structure of Science', Routledge & Kegan Paul, London, 1961, p. 90.

19 Peter Achinstein, 'Concepts of Science', op. cit., p. 203. For other recent philosophical discussions of 'model' see: R. Harré, 'The Principles of Scientific Thinking', Macmillan, London, 1970, chapter 2; Mary Hesse, An Inductive Logic of Theories, in Michael Radner and Steven Winokur (eds), op. cit., and her 'Models and Analogies in Science', Sheed & Ward, London, 1963; W.H. Leatherdale, 'The Role of Analogy, Model and Metaphor in Science', North Holland Publishing Company, Amsterdam, Oxford, 1974. For a stimulating discussion related more directly to sociology see David Willer, 'Scientific Sociology', Prentice-Hall, Englewood Cliffs, 1967, chapters 2 and 3.
20 Hilary Putnam, op. cit., p. 241.
21 W.V. Quine, Two Dogmas of Empiricism, in his 'From a Logical Point of View', Harper & Row, New York, 1963, p. 20.
22 For more detail see Suppe's account of the 'final version' in Frederick Suppe, op. cit., pp. 50-2.
23 For his more extended discussion see W.V. Quine, 'Word and Object', MIT Press, Cambridge (Mass.), 1960.
24 W.V.O. Quine, Two Dogmas of Empiricism, op. cit.; Hilary Putnam, The Analytic and the Synthetic, in Herbert Feigl and Grover Maxwell (eds), 'Minnesota Studies in the Philosophy of Science', vol. III, University of Minnesota Press, Minneapolis, 1962. See also W.V. Quine, Carnap and Logical Truth, in Paul Arthur Schilpp (ed.), 'The Philosophy of Rudolf Carnap', Cambridge University Press, London, 1963; Peter Achinstein, op. cit., pp. 39-46; Jonathan Bennett, 'analytic-synthetic', 'Proceedings of the Aristotelian Society', 59, 1959; and many others.
25 Carnap, of course, wrote voluminously, but Rudolf Carnap, 'Meaning and Necessity', University of Chicago Press, 1947, is probably of most relevance here. There is a second enlarged edition published in 1956. See also Carnap and other's contributions to Paul Arthur Schilpp (ed.), op. cit.
26 Taken primarily from W.V.O. Quine, Two Dogmas of Empiricism, op. cit.
27 For extensive discussion see W.V.O. Quine, 'Word and Object', op. cit.
28 W.V.O. Quine, Two Dogmas of Empiricism, op. cit., p. 42.
29 Hilary Putnam, The Analytic and the Synthetic, op. cit.

30 Collected in Michael Radner and Stephen Winokur (eds), op. cit.
31 Carl G. Hempel, in ibid., pp. 158-61.
32 W.V.O. Quine, Carnap and Logical Truth, op. cit., p. 406.
33 Mary Hesse has made one attempt to offer an account of scientific inference that takes Quine's 'network' model seriously. See Mary Hesse, 'The Structure of Scientific Inference', Macmillan, London, 1974.

CHAPTER 4 ELEMENTS OF THEORY

1 See Max Black, 'Models and Metaphors: Studies in Language and Philosophy', Cornell University Press, Ithaca, NY, 1962, and Mary B. Hesse, 'Models and Analogies in Science', University of Notre Dame Press, Indiana, 1966.
2 Frederick Suppe (ed.), 'The Structure of Scientific Theories', University of Illinois Press, Urbana, Chicago, London, 1974, pp. 125-221. He pays far more attention to this group than to any other of his 'alternatives to the Received View'.
3 Mary B. Hesse, op. cit., pp. 174-5.
4 See R. Harré, 'Theories and Things', Sheed & Ward, London, New York, 1961, especially pp. 18-42. See also R. Harré, 'The Principles of Scientific Thinking', Macmillan, London, 1970, particularly chapter 2.
5 Herbert Feigl, Some Remarks on the Meaning of Scientific Explanation, in H. Feigl and W. Sellars (eds), 'Readings in Philosophical Analysis', Appleton-Century-Crofts, New York, 1949. Feyerabend describes this imagery as a layer-cake, a description apparently less to Feigl's taste than to mine; see Herbert Feigl, The 'Orthodox' View of Theories: Remarks in Defence as well as Critique, in Michael Radner and Stephen Winokur (eds), 'Minnesota Studies in the Philosophy of Science', vol. IV, University of Minnesota Press, Minneapolis, 1970.
6 For one Parsonian account see Talcott Parsons, 'The Structure of Social Action', Free Press, Glencoe, Ill., 1949, and for rather different emphases see Talcott Parsons, 'The Social System', Routledge & Kegan Paul, London, 1951. For a useful brief comment on relevant continuities and disjunctions in Parsonian action theory

see Robert Dubin, Parsons' Actor: Continuities in Socia Social Theory, 'American Sociological Review', 25, 4, 1960. For Schutz's initial development of Weber's work see Alfred Schutz, 'The Phenomenology of the Social World', Heinemann, London, 1972. Dahrendorf developed his views in Ralf Dahrendorf, 'Class and Class Conflict in Industrial Society', Routledge & Kegan Paul, London, 1959, and Rex in John Rex, 'Key Problems of Sociological Theory', Routledge & Kegan Paul, London, 1961.
7 See Max Black, op. cit.; Mary B. Hesse, op. cit.; R. Harré, 'Theories and Things', op. cit. For an extensive summary of the literature see W.H. Leatherdale, 'The Role of Analogy, Model and Metaphor in Science', North Holland Publishing Company, Amsterdam, Oxford, 1974. He summarises the range of conceptions of metaphor's role in science on pp. 172-3.
8 See Max Black, op. cit., pp. 30-47. His inspiration for the 'interaction' view is Richards. See I.A. Richards, 'The Philosophy of Rhetoric', Oxford University Press, New York, 1965.
9 R. Harré, 'Theories and Things', op. cit.
10 David Willer and Judith Willer, 'Systematic Empiricism: Critique of a Pseudo-Science', Prentice-Hall, Englewood Cliffs, 1973.
11 Paul Feyerabend, 'Against Method', New Left Books, London, 1975, p. 295.

CHAPTER 5 ON EXPLANATION

1 Carl Hempel, 'Aspects of Scientific Explanation', Free Press, New York, Collier-Macmillan, London, 1965, pp. 345-6.
2 As examples of contributions to the debate see Carl Hempel, The Function of General Laws in History, 'The Journal of Philosophy', 39, 1942; reprinted in a slightly modified version in 'Aspects of Scientific Explanation', op. cit.; William Dray, 'Laws and Explanations in History', Oxford University Press, 1957; Patrick Gardiner, 'The Nature of Historical Explanation', Oxford University Press, 1952; Arthur C. Danto, 'Analytical Philosophy of History', Cambridge University Press, 1968, especially chapter 10; and the materials collected in Patrick Gardiner (ed.), 'Theories of History', Free Press, Glencoe, Ill., 1959.

3 Carl Hempel, The Logic of Functional Analysis, in Llewellyn Gross (ed.), 'Symposium on Sociological Theory', Harper & Row, New York, 1959, reprinted in 'Aspects of Scientific Explanation', op. cit.
4 George C. Homans, 'Social Behaviour: its elementary forms', Harcourt Brace Jovanovich, New York, 1961; 'The Nature of Social Science', Harcourt, Brace & World, New York, 1967.
5 R.B. Braithwaite, 'Scientific Explanation', Harper & Row, New York, 1960 (first published 1953). The arguments with 'Social Behaviour' concern its behavioural account of exchange relations as much as its commitment to a particular view of explanation.
6 George C. Homans, 'The Nature of Social Science', op. cit., p. 23.
7 It would be a major task to document the many ramifications of naive deductivism. For fairly formal and therefore highly visible examples see: H.L. Zetterberg, 'On Theory and Verification in Sociology', Bedminster Press, Totlowa, NJ, 1965; Arthur L. Stinchcombe, 'Constructing Social Theories', Harcourt, Brace & World, New York, 1968. For more openly mathematical views see Joseph Berger, Morris Zelditch, Jr and Bo Anderson (eds), 'Sociological Theories in Progress', vols 1 and 2, Houghton Mifflin, Boston, 1966 and 1973; Peter Abell, 'Model Building in Sociology', Weidenfeld & Nicolson, London, 1971. For a good discussion of the influence of this sort of view of science in sociology see Richard J. Bernstein, 'The Restructuring of Social and Political Theory', Methuen, London, 1976, part I.
8 See, for example, the materials collected in Alan Ryan (ed.), 'The Philosophy of Social Explanation', Oxford University Press, London, 1973.
9 Carl Hempel and Paul Oppenheim, Studies in the Logic of Explanation, 'Philosophy of Science', 15, 1948. Reprinted in 'Aspects of Scientific Explanation', op. cit.
10 Danto makes this point very clearly. See Arthur C. Danto, op. cit., chapter 10. See also Peter Achinstein, 'Law and Explanation', Oxford University Press, New York, 1971.
11 See Carl Hempel, Aspects of Scientific Explanation, op. cit., pp. 338-45. For a concise account of some of the problems of the law concept in explanation see Peter Achinstein, op. cit.
12 For a summary of the various positions see Arthur C.

Danto, op. cit., pp. 203-15. For Hempel's responses see Aspects of Scientific Explanation, op. cit., throughout.

13. Michael Scriven, Definitions, Explanations, and Theories, in 'Minnesota Studies in the Philosophy of Science', vol. II, University of Minnesota Press, Minneapolis, 1958, p. 193.
14. Mario Bunge, 'Method, Model and Matter', Reidel, Dordrecht, 1973.
15. Michael Scriven, Truisms as Grounds for Historical Explanations, in Patrick Gardiner (ed.), 'Theories of History', op. cit., p. 467.
16. Carl Hempel, Aspects of Scientific Explanation, op. cit., p. 398.
17. For a useful discussion of explanations employing probabilistic 'laws' see Nicholas Rescher, 'Scientific Explanation', Free Press, New York, Collier-Macmillan, London, 1970.
18. Those unfamiliar with the sort of generalisations I have in mind could obtain some idea by consulting Bernard Berelson and G.A. Steiner, 'Human Behaviour: An Inventory of Scientific Findings', Harcourt, Brace & World, New York, 1964.
19. Carl Hempel, Aspects of Scientific Explanation, op. cit., p. 379.
20. Ibid., p. 440.
21. I am not thinking of any studies in particular here, though Brian Jackson and Denis Marsden, 'Education and the Working Class', Routledge & Kegan Paul, London, 1962, was clearly important in the development of such views.
22. This more holistic conception reflects in questions of explanation a similar claim to that developed by Quine in relation to 'testing', a view which has influenced philosophy of science variously, particularly in the development of the so-called 'network model'. See, for example, Mary Hesse, 'The Structure of Scientific Inference', Macmillan, London, 1974. For a careful attempt to relate this more holistic, post-empiricist philosophy of science to the social sciences see David Thomas, 'Naturalism and Social Science', Cambridge University Press, 1979.
23. As reported, for example, in Basil Bernstein, 'Class, Codes and Control', vol. 1, Routledge & Kegan Paul, London, 1971.

24 For one account see Mario Bunge, 'Causality', Harvard University Press, Cambridge (Mass.), 1959.
25 This is true (covertly) of even the most seemingly 'non-theoretical' of approaches to imputing causal relations on the basis of probabilistic materials. See Hubert M. Blalock, 'Causal Inference in Non-Experimental Research', University of North Carolina Press, Chapel Hill, 1964, and Hubert M. Blalock, Theory Building and Causal Inferences, in Hubert M. Blalock and Ann B. Blalock (eds), 'Methodology in Social Research', McGraw-Hill, New York, London, 1968.
26 This is drawn from Harré's concern with 'ontological experiments'. See R. Harré, 'Theories and Things', Sheed & Ward, London, New York, 1961, particularly pp. 67-71.
27 For a useful summary of work on models in science see W.H. Leatherdale, 'The Role of Analogy, Model and Metaphor in Science', North Holland Publishing Company, Amsterdam, Oxford, 1974. This book includes an extensive classified bibliography. See also references listed in chapter 4.
28 Mary Hesse, 'Models and Analogies in Science', University of Notre Dame Press, Indiana, 1966, p. 157.
29 Peter Winch, 'The Idea of a Social Science', Routledge & Kegan Paul, London, 1958. Winch's study provided the occasion for a long-running debate some idea of which can be obtained by consulting the materials collected in Bryan Wilson (ed.), 'Rationality', Blackwell, Oxford, 1970. The debate is still very much alive though it has long since turned outward from the detail of Winch's position. As an example of a recent discussion of some of these issues see Stephen P. Turner, 'Sociological Explanation as Translation', Cambridge University Press, Cambridge, London, New York, 1980. For a concise summary of what is involved see Richard J. Bernstein, 'The Restructuring of Social and Political Theory', op. cit., part II.

CHAPTER 6 ON DEMONSTRATION

1 The distinction between 'dogmatic' and 'sophisticated' falsificationism comes from Lakatos. See, particularly, Imre Lakatos, Falsification and the Methodology of Research Programmes, in Imre Lakatos and Alan Musgrave

(eds), 'Criticism and the Growth of Knowledge', Cambridge University Press, 1970.
2 Karl Popper, 'Conjectures and Refutations', Routledge & Kegan Paul, London, 1974, p. 41.
3 Karl Popper, 'The Logic of Scientific Discovery', Hutchinson, London, 1968 (first English edition 1959); 'Conjectures and Refutations', op. cit. (first edition 1963); 'Objective Knowledge', Oxford University Press, 1972.
4 Stefan Amsterdamski, 'Between Experience and Metaphysics', Reidel, Dordrecht, Boston, 1975, pp. 100-2; Imre Lakatos, Falsification and the Methodology of Research Programmes, op. cit., p. 181.
5 Karl Popper, 'Conjectures and Refutations', op. cit., p. 34. His concern about the 'pseudo-scientific' features of Marx, Freud and others stayed with him. See his 'The Poverty of Historicism', Routledge & Kegan Paul, London, 1960 and 'The Open Society and its Enemies', vol. 2, Routledge & Kegan Paul, London, 1962.
6 Karl Popper, 'Conjectures and Refutations', op. cit., pp. 52-3.
7 Karl Popper, 'The Logic of Scientific Discovery', op. cit., pp. 32-3.
8 Ibid., p. 41.
9 See especially Imre Lakatos, Falsification and the Methodology of Research Programmes, op. cit., pp. 95-103.
10 For a blunt statement on intersubjectivity see 'The Logic of Scientific Discovery', op. cit., p. 44. For claims that science is the 'search for truth' (in Tarski's sense of 'truth') see 'Conjectures and Refutations', op. cit., pp. 228-31.
11 Karl Popper, 'The Logic of Scientific Discovery', op. cit., p. 42.
12 Ibid., p. 80.
13 See Karl Popper, 'Conjectures and Refutations', op. cit., p. 37, and at considerable length both 'The Poverty of Historicism' and 'The Open Society and its Enemies', op. cit.
14 See Karl Popper, 'The Logic of Scientific Discovery', op. cit., chapter 2, section 11. Also see Imre Lakatos, Falsification and the Methodology of Research Programmes, op. cit., pp. 104-6, on the distinction between conservative and revolutionary conventionalism.
15 Karl Popper, 'The Logic of Scientific Discovery', op. cit., p. 76.

16 Pierre Duhem, 'The Aim and Structure of Physical Theory', Princeton University Press, 1954, p. 216.
17 Karl Popper, 'Conjectures and Refutations', op. cit., p. 239.
18 Sandra G. Harding, Introduction, in Sandra G. Harding (ed.), 'Can Theories be Refuted?', Reidel, Dordrecht, Boston, 1976, p. xv.
19 W.V.O. Quine, Two Dogmas of Empiricism, in his 'From a Logical Point of View', Harper & Row, New York, 1963, p. 43.
20 Ibid., p. 41. Note that Quine attributes this argument to Duhem, op. cit., pp. 303-28. For claims that there are important differences between Duhem's and Quine's position see the papers collected in Sandra G. Harding, 'Can Theories be Refuted?', op. cit.
21 Karl Popper, 'The Logic of Scientific Discovery', op. cit., pp. 37-8.
22 Imre Lakatos, Falsification and the Methodology of Research Programmes, op. cit., p. 113.
23 Karl Popper, 'The Logic of Scientific Discovery', op. cit., p. 43.
24 The key sections of 'The Logic of Scientific Discovery', are 7, 8, 21, 22, 23, 25, 27, 28, 29, and 30.
25 Karl Popper, 'The Logic of Scientific Discovery', op. cit., p. 106.
26 Ibid., pp. 93-4, 104-5.
27 Ibid., p. 94.
28 Ibid., p. 105.
29 Ibid., p. 103.
30 Ibid., p. 103.
31 Ibid., p. 103.
32 Ibid., p. 106.
33 Ibid., p. 108.
34 Ibid., chapter VI. See also the clear summary in Karl Popper, 'Conjectures and Refutations', op. cit., pp. 217-20.
35 Karl Popper, 'The Logic of Scientific Discovery', op. cit., p. 267.
36 Ibid., p. 275.
37 For a relatively short account see Karl Popper, Truth, Rationality and the Growth of Knowledge, in 'Conjectures and Refutations', op. cit., pp. 215-50.
38 Paul Feyerabend, 'Science in a Free Society', New Left Books, London, 1978, p. 7. See also Paul Feyerabend, 'Against Method', New Left Books, London, 1975.

39 See, among others, Paul Feyerabend, 'Against Method', op. cit.; Consolations for the Specialist, in Imre Lakatos and Alan Musgrave (eds), 'Criticism and the Growth of Knowledge', op. cit.; On the Critique of Scientific Reason, in R.S. Cohen, P.K. Feyerabend and M.W. Wartofsky (eds), 'Essays in Memory of Imre Lakatos', Reidel, Dordrecht, Boston, 1976; In Defence of Aristotle: Comments on the Condition of Content Increase, in Gerard Radnitzky and Gunnar Andersson (eds), 'Progress and Rationality in Science', Reidel, Dordrecht, Boston, 1978. See also Thomas Kuhn, 'The Structure of Scientific Revolutions', University of Chicago Press, 1962; and his essays collected in 'The Essential Tension', University of Chicago Press, London, 1977.

40 There has been a thread of such criticism. In recent years the most persistent analysis of this and related problems is to be found in Grunbaum's work. See particularly Adolf Grunbaum, Is Falsifiability the Touchstone of Scientific Rationality? Karl Popper versus Inductivism, in R.S. Cohen, P.K. Feyerabend and M.W. Wartofsky (eds), 'Essays in Memory of Imre Lakatos', op. cit., a selection from which appears as Popper vs Inductivism, in Gerard Radnitzky and Gunnar Andersson (eds), 'Progress and Rationality in Science', op. cit. Further references are listed there.

41 Imre Lakatos, Falsification and the Methodology of Research Programmes, op. cit., p. 103. See also Imre Lakatos, History of Science and its Rational Reconstructions, in Roger C. Buck and Robert S. Cohen (eds), 'Proceedings of the 1970 Biennial Meeting of the Philosophy of Science Association', Reidel, Dordrecht, 1971.

42 Imre Lakatos, Falsification and the Methodology of Research Programmes, op. cit., p. 107. It is arguable that this formulation presumes an answer to the question at issue: how do we decide which theories are successful?

43 Ibid., p. 116.

44 John Worrall, The Ways in which the Methodology of Scientific Research Programmes Improves on Popper's Methodology, in Gerard Radnitzky and Gunnar Andersson (eds), 'Progress and Rationality in Science', op. cit., pp. 55-7.

45 Imre Lakatos, Falsification and the Methodology of Research Programmes, op. cit., p. 129.

46 Ibid., p. 130.
47 For the reference to Kuhn see ibid., p. 132, and for the general exposition see pp. 132-8.
48 Ibid., p. 135.
49 For Kuhn's views see his 'The Structure of Scientific Revolutions', op. cit., and Reflections on my Critics, in Imre Lakatos and Alan Musgrave (eds), 'Criticism and the Growth of Knowledge', op. cit., pp. 266-77. In the same volume see Paul Feyerabend, Consolations for the Specialist especially pp. 219-29 and 'Against Method', op. cit., especially chapters 15, 16, and 17.
50 Imre Lakatos, Falsification and the Methodology of Research Programmes, op. cit., p. 118.

CHAPTER 7 METHODOLOGICAL DIVERSITY IN SOCIOLOGY

1 As recent attempts to explore what might be involved in maintaining a naturalistic view of the social sciences see David Papineau, 'For Science in the Social Sciences', Macmillan, London, 1978; David Thomas, 'Naturalism and Social Science', Cambridge University Press, 1979; Roy Bhaskar, 'The Possibility of Naturalism', Harvester Press, Brighton, 1979. Bhaskar's case is the most original, and I shall return to it later in this chapter.
2 For example: Don Martindale, 'The Nature and Types of Sociological Theory', Routledge & Kegan Paul, London, 1960; Pitirim A. Sorokin, 'Sociological Theories of Today', Harper & Row, New York, Evanston, London 1966; Walter L. Wallace, Overview of Contemporary Sociological Theory, in Walter L. Wallace (ed.), 'Sociological Theory', Heinemann, London, 1969; Roland Robertson, Towards the Identification of the Major Axes of Sociological Analysis, in John Rex (ed.), 'Approaches to Sociology', Routledge & Kegan Paul, London, Boston, 1974; and many more too numerous to mention.
3 Peter Winch, 'The Idea of a Social Science', Routledge & Kegan Paul, London, 1958. Though I do not accept Winch's case this does not mean that I would wish to dismiss the whole tradition which takes as its focus the nature of social action. For the purposes of this discussion, however, I am bracketing that topic.
4 Hans Reichenbach, 'Experience and Prediction', University of Chicago Press, 1938. For a not untypical modern

use of such terms in relation to sociology see Michael Phillipson, Theory, Methodology and Conceptualisation, in Paul Filmer, Michael Phillipson, David Silverman and David Walsh, 'New Directions in Sociological Theory', Collier-Macmillan, London, 1972. For a critical discussion see Andrew Tudor, Misunderstanding Everyday Life, 'The Sociological Review', 24, 3, 1976.
5 H.L. Zetterberg, 'On Theory and Verification in Sociology', Bedminster Press, Totlowa, NJ, 1965. The work of the so-called Stanford School has been influenced by Zetterberg's formulation; see Joseph Berger, Morris Zelditch, Jr and Bo Anderson (eds), 'Sociological Theories in Progress', vols 1 and 2, Houghton Mifflin, Boston, 1966 and 1973. For a more sophisticated case than Zetterberg's, and one which is concerned to develop models formulable in precise propositional terms, see Peter Abell, 'Model Building in Sociology', Weidenfeld & Nicolson, London, 1971. Note that Abell's sense of 'model' is rather different from that of the present study; model building, for him, involves arranging concepts into 'sets of interrelated propositions'. For a case in support of the use of mathematics in relation to sociological propositions see Patrick Doreian, 'Mathematics and the Study of Social Relations', Weidenfeld & Nicolson, London, 1970.
6 George C. Homans, 'Social Behaviour', Routledge & Kegan Paul, London, 1961, p. 10.
7 Note, however, that this is a formal resemblance; there are important contrasts, not the least of which lies in the fact that codes are conceived to 'govern' systems of differences not the determinate associations characteristic of 'law-like' statements. For Saussure's original formulations see Ferdinand de Saussure, 'Course in General Linguistics', Fontana/Collins, London, 1974.
8 See, for example, the style of 'theory' offered in Talcott Parsons, 'The Social System', Routledge & Kegan Paul, London, 1951, which he suggests is 'a theoretical work in a strict sense' and which 'though ultimately to be tested in terms of its usefulness in empirical research' is none the less offered without such assessment. This, Parsons says, 'will have to be undertaken separately' (p. 3). See also Talcott Parsons and Neil J. Smelser, 'Economy and Society', Routledge & Kegan Paul, London, 1956; and the essays collected in Talcott Parsons, 'Sociological Theory and Modern Society', Mac-

millan, London, New York, 1967. For discussion of a non-Parsonian deductive-holist approach see Walter Buckley, 'Sociology and Modern Systems Theory', Prentice-Hall, Englewood Cliffs, 1967.
9 David Willer and Judith Willer, 'Systematic Empiricism: Critique of a Pseudo-Science', Prentice-Hall, Englewood Cliffs, 1973.
10 It seems to me that many of the 'new' sociologies of the early 1970s, though radically different from traditional sociological practice in some respects, remain systematically empiricist. This is particularly true of those who claimed that we should 'stay close to the phenomena', grounding our work in such detailed observation. As an example of a general statement of this view see Jack D. Douglas, Understanding Everyday Life, in Jack D. Douglas (ed.), 'Understanding Everyday Life', Routledge & Kegan Paul, London, 1971. For an argument that such approaches represent more or less esoteric variations on inductivism see Andrew Tudor, Misunderstanding Everyday Life, op. cit.
11 See Peter L. Berger and Thomas Luckmann, 'The Social Construction of Reality', Allen Lane, London, 1967; Peter L. Berger, 'The Social Reality of Religion', Faber & Faber, London, 1969; Peter L. Berger, Brigitte Berger and Hansfried Kellner, 'The Homeless Mind', Penguin, Harmondsworth, 1974.
12 See Erving Goffman, 'The Presentation of Self in Everyday Life', Penguin, Harmondsworth, 1971; 'Stigma', Prentice-Hall, Englewood Cliffs, 1964; 'Encounters', Bobbs-Merrill, Indianapolis, 1961; 'Relations in Public', Harper & Row, New York, 1972; 'Frame Analysis', Penguin, Harmondsworth, 1975. The charge that Goffman's work is somehow not sociological is surely misdirected, though it could be argued that as a sociology it is incomplete and unclear about its own methodological status. But it is hard to generalise. The Goffman of 'Frame Analysis' is rather different from the Goffman of 'Relations in Public'.
13 See, among others, Hubert M. Blalock, 'Causal Inference in Nonexperimental Research', University of North Carolina Press, Chapel Hill, 1964; Hubert M. Blalock, 'Theory Construction', Prentice-Hall, Englewood Cliffs, 1969.
14 The classic statement of which is to be found in Robert K. Merton, Manifest and Latent Functions, in his 'Social

Theory and Social Structure', Free Press, Glencoe, Ill., 1957.
15 Imre Lakatos, Falsification and the Methodology of Scientific Research Programmes, in Imre Lakatos and Alan Musgrave (eds), 'Criticism and the Growth of Knowledge', Cambridge University Press, 1970, p. 107.
16 See Harold Garfinkel, 'Studies in Ethnomethodology', Prentice-Hall, Englewood Cliffs, 1967. For examples of the impressive range of work inspired by the ethnomethodological orientation see Roy Turner (ed.), 'Ethnomethodology', Penguin, Harmondsworth, 1974.
17 David Thomas, 'Naturalism and Social Science', op. cit., p. 7.
18 Initially in Roy Bhaskar, On the Possibility of Social Scientific Knowledge and the Limits of Naturalism, 'Journal for the Theory of Social Behaviour', 8, 1, 1978, an article reprinted in John Mepham and David-Hillel Ruben (eds), 'Issues in Marxist Philosophy', vol. III, Harvester Press, Brighton, 1979, and incorporated into the extended discussion of Roy Bhaskar, 'The Possibility of Naturalism', op. cit. His position is rooted in his general account of science: Roy Bhaskar, 'A Realist Theory of Science', Leeds Books, Leeds, 1975. Of course his is not the only contemporary attempt to reconsider the naturalist argument in relation to the social sciences. For interesting discussions pursuing the question into the topics of value, truth, and meaning, see Hilary Putnam, 'Meaning and the Moral Sciences', Routledge & Kegan Paul, London, Boston, 1978, and Mary Hesse, Theory and Value in the Social Sciences, in Christopher Hookway and Philip Pettit (eds), 'Action and Interpretation', Cambridge University Press, 1978.
19 Roy Bhaskar, On the Possibility of Social Scientific Knowledge and the Limits of Naturalism, op. cit., p. 6.
20 Ibid., p. 14.
21 For use of these terms see Thomas P. Wilson, Normative and Interpretive Paradigms in Sociology, in Jack D. Douglas (ed.), 'Understanding Everyday Life', op. cit.; Alan Dawe, The Two Sociologies, 'British Journal of Sociology', 21, 2, 1970, a usage somewhat revised in Alan Dawe, Theories of Social Action, in Tom Bottomore and Robert Nisbet (eds), 'A History of Sociological Analysis', Heinemann, London, 1979; David Walsh, Sociology and the Social World, in Paul Filmer, Michael Phillipson, David Silverman and David Walsh, op. cit.

22 See Anthony Giddens, 'New Rules of Sociological Method', Hutchinson, London, 1976; 'Studies in Social and Political Theory', Hutchinson, London, 1977; 'Central Problems in Social Theory', Macmillan, London, 1979.

Name index

Abell, P., 194n, 201n
Achinstein, P., 58, 190n, 191n, 194n
Adler, A., 122, 123
Amsterdamski, S., 121, 185n, 189n, 197n
Anderson, B., 194n, 201n
Andersson, G., 185n, 188n, 189n, 199n
Aune, B., 186n
Ayer, A.J., 186n

Bennett, J., 191n
Benton, T., 183n
Berelson, B., 195n
Berger, B., 202n
Berger, J., 194n, 201n
Berger, P., 169, 188n, 202n
Bernstein, B., 110, 195n
Bernstein, R., 157, 183n, 194n, 196n
Bhaskar, R., 178, 179, 180, 181, 200n, 203n
Black, M., 73, 74, 82, 192n, 193n
Blalock, A., 196n
Blalock, H.M., 171, 196n, 202n
Bottomore, T., 184n, 203n
Braithwaite, R., 93, 194n

Buck, R., 199n
Buckley, W., 202n
Bunge, M., 18, 101, 195n, 196n

Carnap, R., 50, 58, 59, 64, 67, 104, 191n
Cohen, R.S., 186n, 189n, 199n
Colodny, R., 186n, 189n

Dahrendorf, R., 81, 193n
Danto, A., 21, 186n, 187n, 189n, 193n, 194n, 195n
Dawe, A., 203n
Descartes, R., 41
Doreian, P., 201n
Douglas, J., 202n, 203n
Dray, W., 92, 193n
Dubin, R., 193n
Duhem, P., 126, 128, 129, 130, 140, 146, 147, 198n
Durkheim, E., 163

Einstein, A., 122
Emmet, D., 184n
Evans-Pritchard, E.E., 188n

Feigl, H., 78, 190n, 191n, 192n
Feyerabend, P., 17, 32, 33, 34, 35, 36, 39, 47, 75, 87, 114, 140, 141, 149, 185n, 186n, 187n, 188n, 192n, 193n, 198n, 199n, 200n
Filmer, P., 201n, 203n
Freud, S., 83, 122, 123, 125, 197n
Fries, J.F., 133

Gardiner, P., 193n, 195n
Garfinkel, H., 203n
Giddens, A., 181, 184n, 204n
Goodman, N., 16, 17, 19, 185n
Goffman, E., 81, 169, 202n
Gouldner, A., 186n
Grunbaum, A., 199n

Hamlyn, D.W., 26, 27, 186n, 187n
Hanson, A.H., 186n
Harding, S., 129, 198n
Harre, R., 83, 91, 184n, 189n, 190n, 191n, 192n, 193n, 196n
Hempel, C., 40, 49, 50, 51, 52, 53, 57, 70, 86, 89, 92, 93, 94, 95, 97, 98, 99, 100, 101, 103, 104, 105, 107, 111, 113, 116, 119, 185n, 186n, 188n, 189n, 190n, 193n, 194n, 195n
Hesse, M., 74, 76, 113, 114, 191n, 192n, 193n, 195n, 196n, 203n
Hindess, B., 183n
Holzner, B., 23, 187n

Homans, G., 93, 94, 164, 165, 190n, 194n, 201n
Hookway, C., 203n
Hume, D., 9, 122, 123

Jackson, B., 195n

Kant, I., 122, 123
Keat, R., 183n
Kellner, H., 202n
Kuhn, T., 10, 17, 30, 31, 32, 33, 35, 47, 49, 50, 114, 141, 149, 150, 156, 184n, 185n, 187n, 199n, 200n

Lakatos, I., 32, 35, 121, 131, 139, 140, 144, 145, 146, 147, 148, 149, 150, 151, 153, 154, 156, 175, 185n, 186n, 187n, 189n, 196n, 197n, 198n, 199n, 200n, 203n
Leatherdale, W.H., 191n, 193n, 196n
Lessnoff, M., 183n
Locke, J., 9
Luckmann, T., 202n
Lurie, A., 8, 23, 184n, 187n

MacIntyre, A., 184n
Marsden, D., 195n
Martindale, D., 200n
Marx, K., 122, 123, 125, 128, 197n
Maxwell, G., 191n
Medawar, P., 188n
Mepham, J., 203n
Merton, R.K., 202n
Mill, J.S., 9
Morganbesser, S., 5, 184n, 189n

Musgrave, A., 185n, 186n, 187n, 189n, 196n, 199n, 200n, 203n

Nagel, E., 6, 58, 185n, 189n, 190n
Neurath, O., 177
Nisbet, R., 184n, 203n
Northrop, F.S.C., 190n
Nossiter, T., 186n

Oppenheim, P., 95, 188n, 194n

Papineau, D., 200n
Parsons, T., 1, 164, 165, 166, 173, 174, 192n, 201n
Pettit, P., 203n
Phillips, D.L., 34, 188n
Phillipson, M., 201n, 203n
Popper, K., 21, 89, 94, 119, 120, 121, 122, 123, 124, 125, 126, 127, 128, 129, 130, 131, 132, 133, 134, 135, 136, 137, 138, 139, 140, 141, 142, 143, 144, 145, 146, 147, 148, 151, 152, 153, 154, 155, 156, 185n, 187n, 188n, 197n, 198n, 199n
Putnam, H., 12, 49, 51, 53, 58, 60, 65, 66, 68, 69, 77, 84, 114, 162, 189n, 190n, 191n, 203n

Quine, W.V.O., 14, 20, 41, 42, 59, 60, 61, 64, 65, 66, 67, 68, 69, 70, 77, 84, 114, 118, 119, 128, 129, 130, 141, 142, 143, 146, 147, 162, 184n, 186n, 187n, 188n, 191n, 192n, 195n, 198n

Radner, M., 189n, 190n, 191n, 192n
Radnitzky, G., 185n, 188n, 189n, 199n
Reichenbach, H., 161, 200n
Rescher, N., 195n
Rex, J., 81, 184n, 190n, 193n, 200n
Richards, I.A., 82, 193n
Robertson, R., 186n, 200n
Rokkan, S., 186n
Ruben, D.-H., 203n
Rudner, R., 183n
Ryan, A., 183n, 194n
Ryle, G., 19, 186n

Saussure, F. de, 164, 201n
Scheffler, I., 9, 184n
Schilpp, P.A., 191n
Schutz, A., 3, 81, 193n
Scriven, M., 99, 101, 102, 117, 195n
Sellars, W., 192n
Silverman, D., 201n, 203n
Smelser, N.J., 201n
Sorokin, P., 200n
Steiner, G.A., 195n
Stinchcombe, A., 194n
Stoppard, T., 29, 187n
Suppe, F., 49, 51, 53, 56, 75, 189n, 190n, 191n
Suppes, P., 189n

Tarski, A., 139, 189n, 197n
Thomas, D., 195n, 200n, 203n
Toulmin, S., 45, 46, 47, 49, 75, 189n
Tudor, A., 201n, 202n

Turner, R., 203n
Turner, S., 196n

Ullian, J.S., 188n
Urry, J., 183n

Wallace, W., 183n, 200n
Walsh, D., 201n, 203n
Wartofsky, M.W., 186n, 189n, 199n
Watkins, J., 32, 185n, 187n
Weber, M., 3, 6, 81, 193n
Whitehead, A.N., 165

Willer, D., 8, 84, 167, 183n, 184n, 191n, 193n, 202n
Willer, J., 8, 64, 167, 184n, 193n, 202n
Wilson, B., 196n
Wilson, T., 203n
Winch, P., 116, 159, 183n, 196n, 200n
Winokur, S., 189n, 190n, 191n, 192n

Zelditch, M., 194n, 201n
Zetterberg, H.L., 163, 164, 165, 190n, 194n, 201n

Subject index

abstract modelling, 165-6, 170, 173-4
abstraction, 21, 81-2
action, 11, 81, 114, 116, 176-7, 178-82, 200n; general theory of, 165-6
anarchistic epistemology, 32-5
analytic/synthetic distinction, 18, 54, 64-70, 78
auxiliary hypotheses, 130, 149
auxiliary theories, 135, 175-6
axiomatic calculus, 52, 53, 57, 76, 78, 86, 160, 163
axiomatisation, 54, 76, 163-4

background knowledge, 130, 141, 145
basic statements, 126; as conventions, 132, 136; and corroboration, 137-8, 143; and falsification, 131-7, 145; and objectivity, 132; and observation, 134-6
belief: and knowledge, 21; web of, 40-3
bridge principles, 70, 86

cause: in explanation, 97, 106, 111; in sociology, 171
codes, 164, 201n; restricted and elaborated, 110
coherence, 11, 25, 26, 27, 119
common sense, 20, 22, 25-6, 41
conceptual scheme, 78, 79, 80-1, 83, 174-5
conditional sentence, 96, 98, 103
conventionalism, 126, 134, 146, 147; and falsification, 126-31, 140-3; of observation, 62-3; revolutionary, 145
conventionalist stratagems, 127, 131, 132, 134, 147
correspondence, 11, 24, 25, 26, 27, 66
correspondence rules, 25, 35, 53, 54, 70, 77, 78, 85, 86
corroboration, 126, 129; and falsification, 137-9, 140, 141-2; and scientific progress, 142-3, 146-7, 148
counterinduction, 34, 39
covering laws, 92, 94-107, 172-3, 186n

critical rationalism, 128, 131, 140, 141, 144, 149, 151

deductive-propositional, 163-5, 172-3
deductive-holism, 165-6, 173-4
deductive structure, 52, 54, 57, 76, 83, 91-2, 106, 161, 172
deductivism, 11, 40, 46, 91-4, 101, 161; and metaphor, 113; in sociology, 92-4, 164-5
demarcation, 121-2, 139, 140, 151; and conventionalism, 130-1; and falsification, 124-6, 130-1; and induction, 122-3
demonstration, 24, 25, 31, 35, 37, 40, 84, 86, 87, 88, 89, 114, 118-56, 160; and basic statements, 131-7; and conventionalism, 126-31, 140-3; and corroboration, 137-9; and empiricism, 118-19; and falsification, 120-7; and scientific progress, 139-51
demystification, 5, 184n
described phenomena, 37-8, 86, 167, 172, 173, 175, 182

empirical content, 137-8
empirical generalisation, 84, 85, 106, 167, 169, 172, 173
empiricism, 4, 9, 10, 46, 61, 73-4, 158, 171; and analytic/synthetic distinction, 64-8; and demonstration, 118-19, 138, 154-5; and observation, 59, 60, 62; and psychologism, 133; and sociology, 8, 9, 50-1, 154-5; systematic, 8, 9, 85, 166-8, 170-2, 202n
epistemic ambiguity, 103, 105
epistemic community, 23, 24, 26, 27, 29, 36, 63, 80, 86, 87, 88, 187n
epistemology, 18, 19, 20-30; anarchistic, 32-5; and sociology, 7, 14, 15, 19, 20
ethnomethodology, 155, 167, 176
experience, 28, 160, 169, 175
explanandum, 95, 96, 100, 108, 112, 114
explanans, 95, 100, 114
explanation, 18, 28, 31, 37, 40, 86, 87, 89, 91-117, 119, 160; and causation, 97, 106, 111; covering, 94-107, 108, 109, 111, 112, 113, 115, 172-3; deductive-nomological, 95-100, 101, 102, 103, 105, 106, 107, 111; deductive-statistical, 100; and deductivism, 91-4; historical, 92, 98-9; inductive-statistical, 100-6, 107, 108, 112; and models, 106-13, 114, 115; post hoc, 104-5; and reference classes, 103-4, 105, 107; in sociology, 114-15, 164, 180
explication, 16-17, 19, 50, 52, 54, 56

fallacy of misplaced concreteness, 165

fallibilism, 145, 154
falsification, 11, 120-7, 176; **and basic statements**, 131-7; and conventionalism, 126-31, 140-3; and corroboration, 137-9; and demarcation, 124-6, 130-1; and law-cluster concepts, 69; and scientific growth, 139-51
falsificationism: dogmatic, 119, 125; methodological, 145, 175; naive, 121, 142, 146, 147
formalisation, 54, 57, 76
formal sociology, 163-5, 166, 170, 172-3, 174
Fries trilemma, 133
functionalism, 46, 92, 167, 171-2

hard core, 149, 153
hermeneutics, 177, 180
heuristic: negative, 148-9; positive, 149
holism, 68, 78, 114, 143, 153, 166, 168-9, 175, 195n
holistic/propositional distinction, 162

incommensurability, 34, 36, 149-51, 154
induction, 35, 125, 133, 144, 161, 167, 168; and demarcation, 122-3
inductive/deductive distinction, 161
inductive-holism, 168-9
inductive-propositional, 166-8
inductivism, 40, 144
interpretation, 28-9, 160; of action, 114-15, 175-7, 178; in sociology, 168-9, 175-7
interpretative sociology, 168-9, 170, 174-5, 180
interpretative theory, 148, 153, 175-6, 182
intersubjectivity, 22-9, 30, 112; and testing, 126, 129-30, 135-6, 145

knowledge, 18-29
Kuhn loss, 150

language: of theory, 78-80, 82, 83, 174; and theory-dependence, 88
langue/parole, 166
law-cluster concepts, 68-9, 84
laws, 98, 109, 114-15, 164-5; and causality, 111; covering, 92, 94-107, 122-3, 186n; probabilistic, 94, 99, 100-6; universal, 105, 115
logical empiricism, 3, 47, 49, 60, 113

marxism, 46, 80, 147
mathematical sociology, 164-5
mechanism, 83, 110-11, 112
metaphor, 77; and deductivism, 113; and models, 74-5, 82-4, 113
methodological rules, 86-9, 136, 140, 141, 142, 143, 152, 158, 181; and conventionalism, 127-8, 141
methodology of research programmes, 145-51
models, 53, 54, 58, 73, 149,

153, 154, 160, 171, 172, 173–5; and abstraction, 81–2; and demonstration, 143, 165; in explanation, 106–13; and mechanism, 83, 110–11, 112; and metaphor, 74–5, 82–4, 113; as speculative instruments, 82; in theory, 74–5, 76–7, 78, 79, 82–4, 85, 88
modus tollens, 124, 129

naturalism, 4–8, 17, 157, 177–82
nomic necessity, 102, 111
normal science, 30, 33, 148

objectivity: and intersubjectivity, 126, 132; and basic statements, 132, 136
observation, 36, 51–2, 54, 61–2, 63; and analytic/synthetic, 64–5; and basic statements, 134–6; and interpretative theories, 153–4; and inductive inference, 122–3; language, 35, 38, 39, 60, 63, 70, 73, 85
observational/theoretical distinction, 54, 58–63, 77, 79
ontological depth, 119
ontological testing, 112
operationism, 51, 59

paradigm, 30, 32–3, 36, 38, 149, 150
partial interpretation, 51–2, 54, 58–9, 60, 62, 63, 65, 69, 70, 73, 86
phenomenology, 155, 161, 168, 180

positivism, 3, 4, 5, 8, 10, 136, 179, 180, 181, 184n; and sociology, 4, 5, 161
post-empiricism, 158
probabilistic sentences, 100, 101, 103, 106, 111
propositional form, 163–5, 166–8, 171
propositional/holistic distinction, 162
psychologism, 133, 134, 135

rational reconstruction, 7, 16, 17, 18, 54, 55–6, 57, 89, 93
received view, 49–72, 94, 160, 162, 163; and analytic/synthetic, 64–70; and deductivism, 91; and models, 76–7, 84; and observational/theoretical, 58–63; problems of, 75–8; and theory, 73–5
reductionism, 59, 60, 69
relativism, 10, 18, 19, 32, 33, 34, 35, 36, 41–2

scientific canopy, 36–40, 41, 48, 88
scientific community, 30, 31
scientific growth: and corroboration, 142; and falsification, 139–51
scientific revolutions, 30, 33
semiology, 164–5
sentence systems, 78, 79, 84–5, 88, 106, 110–11, 112, 143, 160, 170, 172, 173, 174, 175
social structure, 166, 179–82
sociology: and abstract modelling, 165–6, 173–4;

and empiricism, 8, 9; and epistemology, 7, 14, 15; and formalisation, 163-5, 172-3; inductive-propositional, 166-8, 170-2; and inductive-holism, 168-9; and interpretation, 168-9, 175-7, 178; methodological variation in, 159-77, 181-2; and philosophy, 1, 2, 10, 11, 12, 14, 15, 16; and philosophy of science, 2, 3, 7, 15, 45-7, 157-8, 177-8, 181-2
structuralism, 165, 166
synonomy, 66-7
systematic empiricism, 8, 9, 85, 166-8, 170-2, 202n
systematic reconstruction, 18, 48, 49, 50, 55-6, 71, 85
systems, 166, 171-2

testing, 28-9, 31, 84, 118-19, 121, 124, 126, 136, 160, 162, 165, 166-7, 169, 180; severity of, 137-9
theoretical terms, 38, 51-2, 58, 60, 63, 73, 74, 76, 77
theoretician's dilemma, 50-1
theory, 37, 73-85, 160-2; components of, 78-85; and models, 74-5; and received view, 73-4; in sociology, 157-82; structure of, 47-8
theory-dependence, 88-9; of observation, 136, 175
theory languages, 60, 63, 70, 79-80
truth, 18-19, 26, 41, 124, 139; and analytic-synthetic, 64-6; coherence theory, 11, 25, 26, 27; correspondence theory, 11, 24, 25, 26, 27, 66; and demonstrability, 24, 165; and intersubjectivity, 22-3, 28, 126; and knowledge, 21-2; and pragmatism, 25-6

verification, 119, 123-4
verificationism, 59
verstehen, 177
Vienna school, 3

Weltanschaungen, 75
world-view, 125, 168

Routledge Social Science Series
Routledge & Kegan Paul   London, Henley and Boston

39 Store Street,
London WC1E 7DD
Broadway House,
Newtown Road,
Henley-on-Thames,
Oxon RG9 1EN
9 Park Street,
Boston, Mass. 02108

Contents

International Library of Sociology 2
General Sociology 2
Foreign Classics of Sociology 2
Social Structure 3
Sociology and Politics 3
Criminology 4
Social Psychology 4
Sociology of the Family 5
Social Services 5
Sociology of Education 5
Sociology of Culture 6
Sociology of Religion 6
Sociology of Art and Literature 6
Sociology of Knowledge 6
Urban Sociology 7
Rural Sociology 7
Sociology of Industry and Distribution 7
Anthropology 8
Sociology and Philosophy 8
International Library of Anthropology 9
International Library of Phenomenology and Moral Sciences 9
International Library of Social Policy 9
International Library of Welfare and Philosophy 10
Library of Social Work 10
Primary Socialization, Language and Education 12
Reports of the Institute of Community Studies 12
Reports of the Institute for Social Studies in Medical Care 13
Medicine, Illness and Society 13
Monographs in Social Theory 13
Routledge Social Science Journals 13
Social and Psychological Aspects of Medical Practice 14

Authors wishing to submit manuscripts for any series in this catalogue should send them to the Social Science Editor, Routledge & Kegan Paul Ltd, 39 Store Street, London WC1E 7DD.
● *Books so marked are available in paperback.*
○ *Books so marked are available in paperback only.*
All books are in metric Demy 8vo format (216 × 138mm approx.) unless otherwise stated.

International Library of Sociology
General Editor John Rex

GENERAL SOCIOLOGY

Barnsley, J. H. The Social Reality of Ethics. *464 pp.*
Brown, Robert. Explanation in Social Science. *208 pp.*
● Rules and Laws in Sociology. *192 pp.*
Bruford, W. H. Chekhov and His Russia. *A Sociological Study. 244 pp.*
Burton, F. and Carlen, P. Official Discourse. *On Discourse Analysis, Government Publications, Ideology. About 140 pp.*
Cain, Maureen E. Society and the Policeman's Role. *326 pp.*
● Fletcher, Colin. Beneath the Surface. *An Account of Three Styles of Sociological Research. 221 pp.*
Gibson, Quentin. The Logic of Social Enquiry. *240 pp.*
Glassner, B. Essential Interactionism. *208 pp.*
Glucksmann, M. Structuralist Analysis in Contemporary Social Thought. *212 pp.*
Gurvitch, Georges. Sociology of Law. *Foreword by Roscoe Pound. 264 pp.*
Hinkle, R. Founding Theory of American Sociology 1881–1913. *About 350 pp.*
Homans, George C. Sentiments and Activities. *336 pp.*
Johnson, Harry M. Sociology: *A Systematic Introduction. Foreword by Robert K. Merton. 710 pp.*
● Keat, Russell and Urry, John. Social Theory as Science. *278 pp.*
Mannheim, Karl. Essays on Sociology and Social Psychology. *Edited by Paul Keckskemeti. With Editorial Note by Adolph Lowe. 344 pp.*
Martindale, Don. The Nature and Types of Sociological Theory. *292 pp.*
● Maus, Heinz. A Short History of Sociology. *234 pp.*
Myrdal, Gunnar. Value in Social Theory: *A Collection of Essays on Methodology. Edited by Paul Streeten. 332 pp.*
Ogburn, William F. and Nimkoff, Meyer F. A Handbook of Sociology. *Preface by Karl Mannheim. 656 pp. 46 figures. 35 tables.*
Parsons, Talcott and Smelser, Neil J. Economy and Society: *A Study in the Integration of Economic and Social Theory. 362 pp.*
Payne, G., Dingwall, R., Payne, J. and Carter, M. Sociology and Social Research. *About 250 pp.*
Podgórecki, A. Practical Social Sciences. *About 200 pp.*
Podgórecki, A. and Łos, M. Multidimensional Sociology. *268 pp.*
Raffel, S. Matters of Fact. *A Sociological Inquiry. 152 pp.*
● Rex, John. Key Problems of Sociological Theory. *220 pp.*
   Sociology and the Demystification of the Modern World. *282 pp.*
● Rex, John. (Ed.) Approaches to Sociology. *Contributions by Peter Abell, Frank Bechhofer, Basil Bernstein, Ronald Fletcher, David Frisby, Miriam Glucksmann, Peter Lassman, Herminio Martins, John Rex, Roland Robertson, John Westergaard and Jock Young. 302 pp.*
Rigby, A. Alternative Realities. *352 pp.*
Roche, M. Phenomenology, Language and the Social Sciences. *374 pp.*
Sahay, A. Sociological Analysis. *220 pp.*
Strasser, Hermann. The Normative Structure of Sociology. *Conservative and Emancipatory Themes in Social Thought. About 340 pp.*
Strong, P. Ceremonial Order of the Clinic. *267 pp.*
Urry, John. Reference Groups and the Theory of Revolution. *244 pp.*
Weinberg, E. Development of Sociology in the Soviet Union. *173 pp.*

FOREIGN CLASSICS OF SOCIOLOGY

● Gerth, H. H. and Mills, C. Wright. From Max Weber: *Essays in Sociology. 502 pp.*

● **Tönnies, Ferdinand.** Community and Association *(Gemeinschaft und Gesellschaft).* |*Translated and Supplemented by Charles P. Loomis. Foreword by Pitirim A. Sorokin. 334 pp.*

SOCIAL STRUCTURE

Andreski, Stanislav. Military Organization and Society. *Foreword by Professor A. R. Radcliffe-Brown. 226 pp. 1 folder.*
Broom, L., Lancaster Jones, F., McDonnell, P. and **Williams, T.** The Inheritance of Inequality. *About 180 pp.*
Carlton, Eric. Ideology and Social Order. *Foreword by Professor Philip Abrahams. About 320 pp.*
Clegg, S. and **Dunkerley, D.** Organization, Class and Control. *614 pp.*
Coontz, Sydney H. Population Theories and the Economic Interpretation. *202 pp.*
Coser, Lewis. The Functions of Social Conflict. *204 pp.*
Crook, I. and **D.** The First Years of the Yangyi Commune. *304 pp., illustrated.*
Dickie-Clark, H. F. Marginal Situation: *A Sociological Study of a Coloured Group. 240 pp. 11 tables.*
Giner, S. and **Archer, M. S.** (Eds) Contemporary Europe: *Social Structures and Cultural Patterns, 336 pp.*
● **Glaser, Barney** and **Strauss, Anselm L.** Status Passage: *A Formal Theory. 212 pp.*
Glass, D. V. (Ed.) Social Mobility in Britain. *Contributions by J. Berent, T. Bottomore, R. C. Chambers, J. Floud, D. V. Glass, J. R. Hall, H. T. Himmelweit, R. K. Kelsall, F. M. Martin, C. A. Moser, R. Mukherjee and W. Ziegel. 420 pp.*
Kelsall, R. K. Higher Civil Servants in Britain: *From 1870 to the Present Day. 268 pp. 31 tables.*
● **Lawton, Denis.** Social Class, Language and Education. *192 pp.*
McLeish, John. The Theory of Social Change: *Four Views Considered. 128 pp.*
● **Marsh, David C.** The Changing Social Structure of England and Wales, 1871–1961. *Revised edition. 288 pp.*
Menzies, Ken. Talcott Parsons and the Social Image of Man. *About 208 pp.*
● **Mouzelis, Nicos.** Organization and Bureaucracy. *An Analysis of Modern Theories. 240 pp.*
● **Ossowski, Stanislaw.** Class Structure in the Social Consciousness. *210 pp.*
● **Podgórecki, Adam.** Law and Society. *302 pp.*
Renner, Karl. Institutions of Private Law and Their Social Functions. *Edited, with an Introduction and Notes, by O. Kahn-Freud. Translated by Agnes Schwarzschild. 316 pp.*
Rex, J. and **Tomlinson, S.** Colonial Immigrants in a British City. *A Class Analysis. 368 pp.*
Smooha, S. Israel: Pluralism and Conflict. *472 pp.*
Wesolowski, W. Class, Strata and Power. *Trans. and with Introduction by G. Kolankiewicz. 160 pp.*
Zureik, E. Palestinians in Israel. *A Study in Internal Colonialism. 264 pp.*

SOCIOLOGY AND POLITICS

Acton, T. A. Gypsy Politics and Social Change. *316 pp.*
Burton, F. Politics of Legitimacy. *Struggles in a Belfast Community. 250 pp.*
Crook, I. and **D.** Revolution in a Chinese Village. *Ten Mile Inn. 216 pp., illustrated.*
Etzioni-Halevy, E. Political Manipulation and Administrative Power. *A Comparative Study. About 200 pp.*
Fielding, N. The National Front. *About 250 pp.*
● **Hechter, Michael.** Internal Colonialism. *The Celtic Fringe in British National Development, 1536–1966. 380 pp.*
Kornhauser, William. The Politics of Mass Society. *272 pp. 20 tables.*

Korpi, W. The Working Class in Welfare Capitalism. *Work, Unions and Politics in Sweden. 472 pp.*

Kroes, R. Soldiers and Students. *A Study of Right- and Left-wing Students. 174 pp.*

Martin, Roderick. Sociology of Power. *About 272 pp.*

Merquior, J. G. Rousseau and Weber. *A Study in the Theory of Legitimacy. About 288 pp.*

Myrdal, Gunnar. The Political Element in the Development of Economic Theory. *Translated from the German by Paul Streeten. 282 pp.*

Varma, B. N. The Sociology and Politics of Development. *A Theoretical Study. 236 pp.*

Wong, S.-L. Sociology and Socialism in Contemporary China. *160 pp.*

Wootton, Graham. Workers, Unions and the State. *188 pp.*

CRIMINOLOGY

Ancel, Marc. Social Defence: *A Modern Approach to Criminal Problems. Foreword by Leon Radzinowicz. 240 pp.*

Athens, L. Violent Criminal Acts and Actors. *104 pp.*

Cain, Maureen E. Society and the Policeman's Role. *326 pp.*

Cloward, Richard A. and **Ohlin, Lloyd E.** Delinquency and Opportunity: *A Theory of Delinquent Gangs. 248 pp.*

Downes, David M. The Delinquent Solution. *A Study in Subcultural Theory. 296 pp.*

Friedlander, Kate. The Psycho-Analytical Approach to Juvenile Delinquency: *Theory, Case Studies, Treatment. 320 pp.*

Gleuck, Sheldon and **Eleanor.** Family Environment and Delinquency. *With the statistical assistance of Rose W. Kneznek. 340 pp.*

Lopez-Rey, Manuel. Crime. *An Analytical Appraisal. 288 pp.*

Mannheim, Hermann. Comparative Criminology: *A Text Book. Two volumes. 442 pp. and 380 pp.*

Morris, Terence. The Criminal Area: *A Study in Social Ecology. Foreword by Hermann Mannheim. 232 pp. 25 tables. 4 maps.*

Rock, Paul. Making People Pay. *338 pp.*

● **Taylor, Ian, Walton, Paul** and **Young, Jock.** The New Criminology. *For a Social Theory of Deviance. 325 pp.*

● **Taylor, Ian, Walton, Paul** and **Young, Jock.** (Eds) Critical Criminology. *268 pp.*

SOCIAL PSYCHOLOGY

Bagley, Christopher. The Social Psychology of the Epileptic Child. *320 pp.*

Brittan, Arthur. Meanings and Situations. *224 pp.*

Carroll, J. Break-Out from the Crystal Palace. *200 pp.*

● **Fleming, C. M.** Adolescence: Its Social Psychology. *With an Introduction to recent findings from the fields of Anthropology, Physiology, Medicine, Psychometrics and Sociometry. 288 pp.*

● The Social Psychology of Education: *An Introduction and Guide to Its Study. 136 pp.*

Linton, Ralph. The Cultural Background of Personality. *132 pp.*

● **Mayo, Elton.** The Social Problems of an Industrial Civilization. *With an Appendix on the Political Problem. 180 pp.*

Ottaway, A. K. C. Learning Through Group Experience. *176 pp.*

Plummer, Ken. Sexual Stigma. *An Interactionist Account. 254 pp.*

● **Rose, Arnold M.** (Ed.) Human Behaviour and Social Processes: *an Interactionist Approach. Contributions by Arnold M. Rose, Ralph H. Turner, Anselm Strauss, Everett C. Hughes, E. Franklin Frazier, Howard S. Becker et al. 696 pp.*

Smelser, Neil J. Theory of Collective Behaviour. *448 pp.*

Stephenson, Geoffrey M. The Development of Conscience. *128 pp.*

Young, Kimball. Handbook of Social Psychology. *658 pp. 16 figures. 10 tables.*

SOCIOLOGY OF THE FAMILY

Bell, Colin R. Middle Class Families: *Social and Geographical Mobility.* 224 pp.
Burton, Lindy. Vulnerable Children. 272 pp.
Gavron, Hannah. The Captive Wife: *Conflicts of Household Mothers.* 190 pp.
George, Victor and **Wilding, Paul.** Motherless Families. 248 pp.
Klein, Josephine. Samples from English Cultures.
  1. Three Preliminary Studies and Aspects of Adult Life in England. 447 pp.
  2. Child-Rearing Practices and Index. 247 pp.
Klein, Viola. The Feminine Character. *History of an Ideology.* 244 pp.
McWhinnie, Alexina M. Adopted Children. *How They Grow Up.* 304 pp.
● **Morgan, D. H. J.** Social Theory and the Family. *About 320 pp.*
● **Myrdal, Alva** and **Klein, Viola.** Women's Two Roles: *Home and Work.* 238 pp. 27 tables.
Parsons, Talcott and **Bales, Robert F.** Family: Socialization and Interaction Process. *In collaboration with James Olds, Morris Zelditch and Philip E. Slater.* 456 pp. 50 figures and tables.

SOCIAL SERVICES

Bastide, Roger. The Sociology of Mental Disorder. *Translated from the French by Jean McNeil.* 260 pp.
Carlebach, Julius. Caring For Children in Trouble. 266 pp.
George, Victor. Foster Care. *Theory and Practice.* 234 pp.
  Social Security: *Beveridge and After.* 258 pp.
George, V. and **Wilding, P.** Motherless Families. 248 pp.
● **Goetschius, George W.** Working with Community Groups. 256 pp.
Goetschius, George W. and **Tash, Joan.** Working with Unattached Youth. 416 pp.
Heywood, Jean S. Children in Care. *The Development of the Service for the Deprived Child. Third revised edition.* 284 pp.
King, Roy D., Ranes, Norma V. and **Tizard, Jack.** Patterns of Residential Care. 356 pp.
Leigh, John. Young People and Leisure. 256 pp.
● **Mays, John.** (Ed.) Penelope Hall's Social Services of England and Wales. 368 pp.
Morris, Mary. Voluntary Work and the Welfare State. 300 pp.
Nokes, P. L. The Professional Task in Welfare Practice. 152 pp.
Timms, Noel. Psychiatric Social Work in Great Britain (1939–1962). 280 pp.
●   Social Casework: *Principles and Practice.* 256 pp.

SOCIOLOGY OF EDUCATION

Banks, Olive. Parity and Prestige in English Secondary Education: a Study in Educational Sociology. 272 pp.
● **Blyth, W. A. L.** English Primary Education. *A Sociological Description.*
  2. Background. 168 pp.
Collier, K. G. The Social Purposes of Education: *Personal and Social Values in Education.* 268 pp.
Evans, K. M. Sociometry and Education. 158 pp.
● **Ford, Julienne.** Social Class and the Comprehensive School. 192 pp.
Foster, P. J. Education and Social Change in Ghana. 336 pp. 3 maps.
Fraser, W. R. Education and Society in Modern France. 150 pp.
Grace, Gerald R. Role Conflict and the Teacher. 150 pp.
Hans, Nicholas. New Trends in Education in the Eighteenth Century. 278 pp. 19 tables.
●   Comparative Education: *A Study of Educational Factors and Traditions.* 360 pp.
● **Hargreaves, David.** Interpersonal Relations and Education. 432 pp.
●   Social Relations in a Secondary School. 240 pp.
  School Organization and Pupil Involvement. *A Study of Secondary Schools.*

- **Mannheim, Karl** and **Stewart, W. A. C.** An Introduction to the Sociology of Education. *206 pp.*
- **Musgrove, F.** Youth and the Social Order. *176 pp.*
- **Ottaway, A. K. C.** Education and Society: An Introduction to the Sociology of Education. *With an Introduction by W. O. Lester Smith. 212 pp.*

  Peers, Robert. Adult Education: *A Comparative Study. Revised edition. 398 pp.*

  Stratta, Erica. The Education of Borstal Boys. *A Study of their Educational Experiences prior to, and during, Borstal Training. 256 pp.*
- **Taylor, P. H., Reid, W. A.** and **Holley, B. J.** The English Sixth Form. *A Case Study in Curriculum Research. 198 pp.*

SOCIOLOGY OF CULTURE

Eppel, E. M. and **M.** Adolescents and Morality: *A Study of some Moral Values and Dilemmas of Working Adolescents in the Context of a changing Climate of Opinion. Foreword by W. J. H. Sprott. 268 pp. 39 tables.*
- **Fromm, Erich.** The Fear of Freedom. *286 pp.*
- The Sane Society. *400 pp.*

Johnson, L. The Cultural Critics. *From Matthew Arnold to Raymond Williams. 233 pp.*

Mannheim, Karl. Essays on the Sociology of Culture. *Edited by Ernst Mannheim in co-operation with Paul Kecskemeti. Editorial Note by Adolph Lowe. 280 pp.*

Merquior, J. G. The Veil and the Mask. *Essays on Culture and Ideology. Foreword by Ernest Gellner. 140 pp.*

Zijderfeld, A. C. On Clichés. *The Supersedure of Meaning by Function in Modernity. 150 pp.*

SOCIOLOGY OF RELIGION

Argyle, Michael and **Beit-Hallahmi, Benjamin.** The Social Psychology of Religion. *256 pp.*

Glasner, Peter E. The Sociology of Secularisation. *A Critique of a Concept. 146 pp.*

Hall, J. R. The Ways Out. *Utopian Communal Groups in an Age of Babylon. 280 pp.*

Ranson, S., Hinings, B. and **Bryman, A.** Clergy, Ministers and Priests. *216 pp.*

Stark, Werner. The Sociology of Religion. *A Study of Christendom.*
  Volume II. *Sectarian Religion. 368 pp.*
  Volume III. *The Universal Church. 464 pp.*
  Volume IV. *Types of Religious Man. 352 pp.*
  Volume V. *Types of Religious Culture. 464 pp.*

Turner, B. S. Weber and Islam. *216 pp.*

Watt, W. Montgomery. Islam and the Integration of Society. *320 pp.*

SOCIOLOGY OF ART AND LITERATURE

Jarvie, Ian C. Towards a Sociology of the Cinema. *A Comparative Essay on the Structure and Functioning of a Major Entertainment Industry. 405 pp.*

Rust, Frances S. Dance in Society. *An Analysis of the Relationships between the Social Dance and Society in England from the Middle Ages to the Present Day. 256 pp. 8 pp. of plates.*

Schücking, L. L. The Sociology of Literary Taste. *112 pp.*

Wolff, Janet. Hermeneutic Philosophy and the Sociology of Art. *150 pp.*

SOCIOLOGY OF KNOWLEDGE

Diesing, P. Patterns of Discovery in the Social Sciences. *262 pp.*

- **Douglas, J. D.** (Ed.) Understanding Everyday Life. *370 pp.*
- **Hamilton, P.** Knowledge and Social Structure. *174 pp.*
- **Jarvie, I. C.** Concepts and Society. *232 pp.*
- **Mannheim, Karl.** Essays on the Sociology of Knowledge. *Edited by Paul Kecskemeti. Editorial Note by Adolph Lowe. 353 pp.*
- **Remmling, Gunter W.** The Sociology of Karl Mannheim. *With a Bibliographical Guide to the Sociology of Knowledge, Ideological Analysis, and Social Planning. 255 pp.*
- **Remmling, Gunter W.** (Ed.) Towards the Sociology of Knowledge. *Origin and Development of a Sociological Thought Style. 463 pp.*
- **Scheler, M.** Problems of a Sociology of Knowledge. *Trans. by M. S. Frings. Edited and with an Introduction by K. Stikkers. 232 pp.*

URBAN SOCIOLOGY

- **Aldridge, M.** The British New Towns. *A Programme Without a Policy. 232 pp.*
- **Ashworth, William.** The Genesis of Modern British Town Planning: *A Study in Economic and Social History of the Nineteenth and Twentieth Centuries. 288 pp.*
- **Brittan, A.** The Privatised World. *196 pp.*
- **Cullingworth, J. B.** Housing Needs and Planning Policy: *A Restatement of the Problems of Housing Need and 'Overspill' in England and Wales. 232 pp. 44 tables. 8 maps.*
- **Dickinson, Robert E.** City and Region: *A Geographical Interpretation. 608 pp. 125 figures.*
  The West European City: *A Geographical Interpretation. 600 pp. 129 maps. 29 plates.*
- **Humphreys, Alexander J.** New Dubliners: *Urbanization and the Irish Family. Foreword by George C. Homans. 304 pp.*
- **Jackson, Brian.** Working Class Community: *Some General Notions raised by a Series of Studies in Northern England. 192 pp.*
- **Mann, P. H.** An Approach to Urban Sociology. *240 pp.*
- **Mellor, J. R.** Urban Sociology in an Urbanized Society. *326 pp.*
- **Morris, R. N.** and **Mogey, J.** The Sociology of Housing. *Studies at Berinsfield. 232 pp. 4 pp. plates.*
- **Mullan, R.** Stevenage Ltd. *About 250 pp.*
- **Rex, J.** and **Tomlinson, S.** Colonial Immigrants in a British City. *A Class Analysis. 368 pp.*
- **Rosser, C.** and **Harris, C.** The Family and Social Change. *A Study of Family and Kinship in a South Wales Town. 352 pp. 8 maps.*
- **Stacey, Margaret, Batsone, Eric, Bell, Colin** and **Thurcott, Anne.** Power, Persistence and Change. *A Second Study of Banbury. 196 pp.*

RURAL SOCIOLOGY

- **Mayer, Adrian C.** Peasants in the Pacific. *A Study of Fiji Indian Rural Society. 248 pp. 20 plates.*
- **Williams, W. M.** The Sociology of an English Village: *Gosforth. 272 pp. 12 figures. 13 tables.*

SOCIOLOGY OF INDUSTRY AND DISTRIBUTION

- **Dunkerley, David.** The Foreman. *Aspects of Task and Structure. 192 pp.*
- **Eldridge, J. E. T.** Industrial Disputes. *Essays in the Sociology of Industrial Relations. 288 pp.*
- **Hollowell, Peter G.** The Lorry Driver. *272 pp.*
- **Oxaal, I., Barnett, T.** and **Booth, D.** (Eds) Beyond the Sociology of Development.

Economy and Society in Latin America and Africa. 295 pp.
Smelser, Neil J. Social Change in the Industrial Revolution: *An Application of Theory to the Lancashire Cotton Industry, 1770–1840. 468 pp. 12 figures. 14 tables.*
Watson, T. J. The Personnel Managers. *A Study in the Sociology of Work and Employment, 262 pp.*

ANTHROPOLOGY

Brandel-Syrier, Mia. Reeftown Elite. *A Study of Social Mobility in a Modern African Community on the Reef. 376 pp.*
Dickie-Clark, H. F. The Marginal Situation. *A Sociological Study of a Coloured Group. 236 pp.*
Dube, S. C. Indian Village. *Foreword by Morris Edward Opler. 276 pp. 4 plates.*
India's Changing Villages: *Human Factors in Community Development. 260 pp. 8 plates. 1 map.*
Fei, H.-T. Peasant Life in China. *A Field Study of Country Life in the Yangtze Valley. With a foreword by Bronislaw Malinowski. 328 pp. 16 pp. plates.*
Firth, Raymond. Malay Fishermen. *Their Peasant Economy. 420 pp. 17 pp. plates.*
Gulliver, P. H. Social Control in an African Society: a Study of the Arusha, Agricultural Masai of Northern Tanganyika. *320 pp. 8 plates. 10 figures.*
Family Herds. *288 pp.*
Jarvie, Ian C. The Revolution in Anthropology. *268 pp.*
Little, Kenneth L. Mende of Sierra Leone. *308 pp. and folder.*
Negroes in Britain. *With a New Introduction and Contemporary Study by Leonard Bloom. 320 pp.*
Tambs-Lyche, H. London Patidars. *About 180 pp.*
Madan, G. R. Western Sociologists on Indian Society. *Marx, Spencer, Weber, Durkheim, Pareto. 384 pp.*
Mayer, A. C. Peasants in the Pacific. *A Study of Fiji Indian Rural Society. 248 pp.*
Meer, Fatima. Race and Suicide in South Africa. *325 pp.*
Smith, Raymond T. The Negro Family in British Guiana: *Family Structure and Social Status in the Villages. With a Foreword by Meyer Fortes. 314 pp. 8 plates. 1 figure. 4 maps.*

SOCIOLOGY AND PHILOSOPHY

Adriaansens, H. Talcott Parsons and the Conceptual Dilemma. *About 224 pp.*
Barnsley, John H. The Social Reality of Ethics. *A Comparative Analysis of Moral Codes. 448 pp.*
Diesing, Paul. Patterns of Discovery in the Social Sciences. *362 pp.*
● **Douglas, Jack D.** (Ed.) Understanding Everyday Life. *Toward the Reconstruction of Sociological Knowledge. Contributions by Alan F. Blum, Aaron W. Cicourel, Norman K. Denzin, Jack D. Douglas, John Heeren, Peter McHugh, Peter K. Manning, Melvin Power, Matthew Speier, Roy Turner, D. Lawrence Wieder, Thomas P. Wilson and Don H. Zimmerman. 370 pp.*
Gorman, Robert A. The Dual Vision. *Alfred Schutz and the Myth of Phenomenological Social Science. 240 pp.*
Jarvie, Ian C. Concepts and Society. *216 pp.*
Kilminster, R. Praxis and Method. *A Sociological Dialogue with Lukács, Gramsci and the Early Frankfurt School. 334 pp.*
● **Pelz, Werner.** The Scope of Understanding in Sociology. *Towards a More Radical Reorientation in the Social Humanistic Sciences. 283 pp.*
Roche, Maurice. Phenomenology, Language and the Social Sciences. *371 pp.*
Sahay, Arun. Sociological Analysis. *212 pp.*
● **Slater, P.** Origin and Significance of the Frankfurt School. *A Marxist Perspective. 185 pp.*

Spurling, L. Phenomenology and the Social World. *The Philosophy of Merleau-Ponty and its Relation to the Social Sciences.* 222 pp.
Wilson, H. T. The American Ideology. *Science, Technology and Organization as Modes of Rationality.* 368 pp.

International Library of Anthropology
General Editor Adam Kuper

● **Ahmed, A. S.** Millennium and Charisma Among Pathans. *A Critical Essay in Social Anthropology.* 192 pp.
 Pukhtun Economy and Society. *Traditional Structure and Economic Development.* About 360 pp.
Barth, F. Selected Essays. *Volume I. About 250 pp.* Selected Essays. *Volume II. About 250 pp.*
Brown, Paula. The Chimbu. *A Study of Change in the New Guinea Highlands.* 151 pp.
Foner, N. Jamaica Farewell. *200 pp.*
Gudeman, Stephen. Relationships, Residence and the Individual. *A Rural Panamanian Community.* 288 pp. 11 plates, 5 figures, 2 maps, 10 tables.
 The Demise of a Rural Economy. *From Subsistence to Capitalism in a Latin American Village.* 160 pp.
Hamnett, Ian. Chieftainship and Legitimacy. *An Anthropological Study of Executive Law in Lesotho.* 163 pp.
Hanson, F. Allan. Meaning in Culture. *127 pp.*
Hazan, H. The Limbo People. *A Study of the Constitution of the Time Universe Among the Aged.* About 192 pp.
Humphreys, S. C. Anthropology and the Greeks. *288 pp.*
Karp, I. Fields of Change Among the Iteso of Kenya. *140 pp.*
Lloyd, P. C. Power and Independence. *Urban Africans' Perception of Social Inequality.* 264 pp.
Parry, J. P. Caste and Kinship in Kangra. *352 pp. Illustrated.*
Pettigrew, Joyce. Robber Noblemen. *A Study of the Political System of the Sikh Jats.* 284 pp.
Street, Brian V. The Savage in Literature. *Representations of 'Primitive' Society in English Fiction, 1858–1920.* 207 pp.
Van Den Berghe, Pierre L. Power and Privilege at an African University. *278 pp.*

International Library of Phenomenology and Moral Sciences
General Editor John O'Neill

Apel, K.-O. Towards a Transformation of Philosophy. *308 pp.*
Bologh, R. W. Dialectical Phenomenology. *Marx's Method.* 287 pp.
Fekete, J. The Critical Twilight. *Explorations in the Ideology of Anglo-American Literary Theory from Eliot to McLuhan.* 300 pp.
Medina, A. Reflection, Time and the Novel. *Towards a Communicative Theory of Literature.* 143 pp.

International Library of Social Policy
General Editor Kathleen Jones

Bayley, M. Mental Handicap and Community Care. *426 pp.*
Bottoms, A. E. and **McClean, J. D.** Defendants in the Criminal Process. *284 pp.*
Bradshaw, J. The Family Fund. *An Initiative in Social Policy.* About 224 pp.

Butler, J. R. Family Doctors and Public Policy. *208 pp.*
Davies, Martin. Prisoners of Society. *Attitudes and Aftercare. 204 pp.*
Gittus, Elizabeth. Flats, Families and the Under-Fives. *285 pp.*
Holman, Robert. Trading in Children. *A Study of Private Fostering. 355 pp.*
Jeffs, A. Young People and the Youth Service. *160 pp.*
Jones, Howard and Cornes, Paul. Open Prisons. *288 pp.*
Jones, Kathleen. History of the Mental Health Service. *428 pp.*
Jones, Kathleen with **Brown, John, Cunningham, W. J., Roberts, Julian** and **Williams, Peter.** Opening the Door. *A Study of New Policies for the Mentally Handicapped. 278 pp.*
Karn, Valerie. Retiring to the Seaside. *400 pp. 2 maps. Numerous tables.*
King, R. D. and **Elliot, K. W.** Albany: Birth of a Prison—End of an Era. *394 pp.*
Thomas, J. E. The English Prison Officer since 1850: *A Study in Conflict. 258 pp.*
Walton, R. G. Women in Social Work. *303 pp.*
● **Woodward, J.** To Do the Sick No Harm. *A Study of the British Voluntary Hospital System to 1875. 234 pp.*

International Library of Welfare and Philosophy
General Editors Noel Timms and David Watson

● **McDermott, F. E.** (Ed.) Self-Determination in Social Work. *A Collection of Essays on Self-determination and Related Concepts by Philosophers and Social Work Theorists. Contributors: F. P. Biestek, S. Bernstein, A. Keith-Lucas, D. Sayer, H. H. Perelman, C. Whittington, R. F. Stalley, F. E. McDermott, I. Berlin, H. J. McCloskey, H. L. A. Hart, J. Wilson, A. I. Melden, S. I. Benn. 254 pp.*
● **Plant, Raymond.** Community and Ideology. *104 pp.*
Ragg, Nicholas M. People Not Cases. *A Philosophical Approach to Social Work. 168 pp.*
● **Timms, Noel** and **Watson, David.** (Eds) Talking About Welfare. *Readings in Philosophy and Social Policy. Contributors: T. H. Marshall, R. B. Brandt, G. H. von Wright, K. Nielsen, M. Cranston, R. M. Titmuss, R. S. Downie, E. Telfer, D. Donnison, J. Benson, P. Leonard, A. Keith-Lucas, D. Walsh, I. T. Ramsey. 320 pp.*
● Philosophy in Social Work. *250 pp.*
● **Weale, A.** Equality and Social Policy. *164 pp.*

Library of Social Work
General Editor Noel Timms

● **Baldock, Peter.** Community Work and Social Work. *140 pp.*
○ **Beedell, Christopher.** Residential Life with Children. *210 pp. Crown 8vo.*
● **Berry, Juliet.** Daily Experience in Residential Life. *A Study of Children and their Care-givers. 202 pp.*
○ Social Work with Children. *190 pp. Crown 8vo.*
● **Brearley, C. Paul.** Residential Work with the Elderly. *116 pp.*
● Social Work, Ageing and Society. *126 pp.*
● **Cheetham, Juliet.** Social Work with Immigrants. *240 pp. Crown 8vo.*
● **Cross, Crispin P.** (Ed.) Interviewing and Communication in Social Work. *Contributions by C. P. Cross, D. Laurenson, B. Strutt, S. Raven. 192 pp. Crown 8vo.*

- **Curnock, Kathleen** and **Hardiker, Pauline.** Towards Practice Theory. *Skills and Methods in Social Assessments. 208 pp.*
- **Davies, Bernard.** The Use of Groups in Social Work Practice. *158 pp.*
- **Davies, Martin.** Support Systems in Social Work. *144 pp.*
- **Ellis, June.** (Ed.) West African Families in Britain. *A Meeting of Two Cultures. Contributions by Pat Stapleton, Vivien Biggs. 150 pp. 1 Map.*
- **Hart, John.** Social Work and Sexual Conduct. *230 pp.*
- **Hutten, Joan M.** Short-Term Contracts in Social Work. *Contributions by Stella M. Hall, Elsie Osborne, Mannie Sher, Eva Sternberg, Elizabeth Tuters. 134 pp.*
- **Jackson, Michael P.** and **Valencia, B. Michael.** Financial Aid Through Social Work. *140 pp.*
- **Jones, Howard.** The Residential Community. *A Setting for Social Work. 150 pp.*
- (Ed.) Towards a New Social Work. *Contributions by Howard Jones, D. A. Fowler, J. R. Cypher, R. G. Walton, Geoffrey Mungham, Philip Priestley, Ian Shaw, M. Bartley, R. Deacon, Irwin Epstein, Geoffrey Pearson. 184 pp.*
- **Jones, Ray** and **Pritchard, Colin.** (Eds) Social Work With Adolescents. *Contributions by Ray Jones, Colin Pritchard, Jack Dunham, Florence Rossetti, Andrew Kerslake, John Burns, William Gregory, Graham Templeman, Kenneth E. Reid, Audrey Taylor. About 170 pp.*
- ○ **Jordon, William.** The Social Worker in Family Situations. *160 pp. Crown 8vo.*
- **Laycock, A. L.** Adolescents and Social Work. *128 pp. Crown 8vo.*
- **Lees, Ray.** Politics and Social Work. *128 pp. Crown 8vo.*
- Research Strategies for Social Welfare. *112 pp. Tables.*
- ○ **McCullough, M. K.** and **Ely, Peter J.** Social Work with Groups. *127 pp. Crown 8vo.*
- **Moffett, Jonathan.** Concepts in Casework Treatment. *128 pp. Crown 8vo.*
- **Parsloe, Phyllida.** Juvenile Justice in Britain and the United States. *The Balance of Needs and Rights. 336 pp.*
- **Plant, Raymond.** Social and Moral Theory in Casework. *112 pp. Crown 8vo.*
- **Priestley, Philip, Fears, Denise** and **Fuller, Roger.** Justice for Juveniles. *The 1969 Children and Young Persons Act; A Case for Reform? 128 pp.*
- **Pritchard, Colin** and **Taylor, Richard.** Social Work: Reform or Revolution? *170 pp.*
- ○ **Pugh, Elisabeth.** Social Work in Child Care. *128 pp. Crown 8vo.*
- **Robinson, Margaret.** Schools and Social Work. *282 pp.*
- ○ **Ruddock, Ralph.** Roles and Relationships. *128 pp. Crown 8vo.*
- **Sainsbury, Eric.** Social Diagnosis in Casework. *118 pp. Crown 8vo.*
- Social Work with Families. *Perceptions of Social Casework among Clients of a Family Service. 188 pp.*
- **Seed, Philip.** The Expansion of Social Work in Britain. *128 pp. Crown 8vo.*
- **Shaw, John.** The Self in Social Work. *124 pp.*
- **Smale, Gerald G.** Prophecy, Behaviour and Change. *An Examination of Self-fulfilling Prophecies in Helping Relationships. 116 pp. Crown 8vo.*
- **Smith, Gilbert.** Social Need. *Policy, Practice and Research. 155 pp.*
- Social Work and the Sociology of Organisations. *124 pp. Revised edition.*
- **Sutton, Carole.** Psychology for Social Workers and Counsellors. *An Introduction. 248 pp.*
- **Timms, Noel.** Language of Social Casework. *122 pp. Crown 8vo.*
- Recording in Social Work. *124 pp. Crown 8vo.*
- **Todd, F. Joan.** Social Work with the Mentally Subnormal. *96 pp. Crown 8vo.*
- **Walrond-Skinner, Sue.** Family Therapy. *The Treatment of Natural Systems. 172 pp.*
- **Warham, Joyce.** An Introduction to Administration for Social Workers. *Revised edition. 112 pp.*
- An Open Case. *The Organisational Context of Social Work. 172 pp.*
- ○ **Wittenberg, Isca Salzberger.** Psycho-Analytic Insight and Relationships. *A Kleinian Approach. 196 pp. Crown 8vo.*

Primary Socialization, Language and Education
General Editor   Basil Bernstein

Adlam, Diana S., *with the assistance of Geoffrey Turner and Lesley Lineker.* Code in Context. *272 pp.*
Bernstein, Basil. Class, Codes and Control. *3 volumes.*
● 1. *Theoretical Studies Towards a Sociology of Language. 254 pp.*
   2. *Applied Studies Towards a Sociology of Language. 377 pp.*
● 3. *Towards a Theory of Educational Transmission. 167 pp.*
Brandis, W. and **Bernstein, B.** Selection and Control. *176 pp.*
Brandis, Walter and **Henderson, Dorothy.** Social Class, Language and Communication. *288 pp.*
Cook-Gumperz, Jenny. Social Control and Socialization. *A Study of Class Differences in the Language of Maternal Control. 290 pp.*
● **Gahagan, D. M.** and **G. A.** Talk Reform. *Exploration in Language for Infant School Children. 160 pp.*
Hawkins, P. R. Social Class, the Nominal Group and Verbal Strategies. *About 220 pp.*
Robinson, W. P. and **Rackstraw, Susan D. A.** A Question of Answers. *2 volumes. 192 pp. and 180 pp.*
Turner, Geoffrey J. and **Mohan, Bernard A.** A Linguistic Description and Computer Programme for Children's Speech. *208 pp.*

Reports of the Institute of Community Studies

Baker, J. The Neighbourhood Advice Centre. A Community Project in Camden. *320 pp.*
● **Cartwright, Ann.** Patients and their Doctors. *A Study of General Practice. 304 pp.*
Dench, Geoff. Maltese in London. *A Case-study in the Erosion of Ethnic Consciousness. 302 pp.*
Jackson, Brian and **Marsden, Dennis.** Education and the Working Class: *Some General Themes Raised by a Study of 88 Working-class Children in a Northern Industrial City. 268 pp. 2 folders.*
Marris, Peter. The Experience of Higher Education. *232 pp. 27 tables.*
●   Loss and Change. *192 pp.*
Marris, Peter and **Rein, Martin.** Dilemmas of Social Reform. *Poverty and Community Action in the United States. 256 pp.*
Marris, Peter and **Somerset, Anthony.** African Businessmen. *A Study of Entrepreneurship and Development in Kenya. 256 pp.*
Mills, Richard. Young Outsiders: *a Study in Alternative Communities. 216 pp.*
Runciman, W. G. Relative Deprivation and Social Justice. *A Study of Attitudes to Social Inequality in Twentieth-Century England. 352 pp.*
Willmott, Peter. Adolescent Boys in East London. *230 pp.*
Willmott, Peter and **Young, Michael.** Family and Class in a London Suburb. *202 pp. 47 tables.*
Young, Michael and **McGeeney, Patrick.** Learning Begins at Home. *A Study of a Junior School and its Parents. 128 pp.*
Young, Michael and **Willmott, Peter.** Family and Kinship in East London. *Foreword by Richard M. Titmuss. 252 pp. 39 tables.*
   The Symmetrical Family. *410 pp.*

Reports of the Institute for Social Studies in Medical Care

Cartwright, Ann, Hockey, Lisbeth and **Anderson, John J.** Life Before Death. *310 pp.*
Dunnell, Karen and **Cartwright, Ann.** Medicine Takers, Prescribers and Hoarders. *190 pp.*
Farrell, C. My Mother Said. . . *A Study of the Way Young People Learned About Sex and Birth Control. 288 pp.*

Medicine, Illness and Society
General Editor W. M. Williams

Hall, David J. Social Relations & Innovation. *Changing the State of Play in Hospitals. 232 pp.*
Hall, David J. and **Stacey, M.** (Eds) Beyond Separation. *234 pp.*
Robinson, David. The Process of Becoming Ill. *142 pp.*
Stacey, Margaret et al. Hospitals, Children and Their Families. *The Report of a Pilot Study. 202 pp.*
Stimson, G. V. and **Webb, B.** Going to See the Doctor. *The Consultation Process in General Practice. 155 pp.*

Monographs in Social Theory
General Editor Arthur Brittan

● **Barnes, B.** Scientific Knowledge and Sociological Theory. *192 pp.*
Bauman, Zygmunt. Culture as Praxis. *204 pp.*
● **Dixon, Keith.** Sociological Theory. *Pretence and Possibility. 142 pp.*
   The Sociology of Belief. *Fallacy and Foundation. About 160 pp.*
Goff, T. W. Marx and Mead. *Contributions to a Sociology of Knowledge. 176 pp.*
Meltzer, B. N., Petras, J. W. and **Reynolds, L. T.** Symbolic Interactionism. *Genesis, Varieties and Criticisms. 144 pp.*
● **Smith, Anthony D.** The Concept of Social Change. *A Critique of the Functionalist Theory of Social Change. 208 pp.*

Routledge Social Science Journals

The British Journal of Sociology. *Editor – Angus Stewart; Associate Editor – Leslie Sklair. Vol. 1, No. 1 – March 1950 and Quarterly. Roy. 8vo. All back issues available. An international journal publishing original papers in the field of sociology and related areas.*
Community Work. *Edited by David Jones and Marjorie Mayo. 1973. Published annually.*
Economy and Society. *Vol. 1, No. 1. February 1972 and Quarterly. Metric Roy. 8vo. A journal for all social scientists covering sociology, philosophy, anthropology, economics and history. All back numbers available.*

Ethnic and Racial Studies. *Editor – John Stone. Vol. 1 – 1978. Published quarterly.*

Religion. Journal of Religion and Religions. *Chairman of Editorial Board, Ninian Smart. Vol. 1, No. 1, Spring 1971. A journal with an inter-disciplinary approach to the study of the phenomena of religion. All back numbers available.*

Sociology of Health and Illness. *A Journal of Medical Sociology. Editor – Alan Davies; Associate Editor – Ray Jobling. Vol. 1, Spring 1979. Published 3 times per annum.*

Year Book of Social Policy in Britain. *Edited by Kathleen Jones. 1971. Published annually.*

Social and Psychological Aspects of Medical Practice
Editor   Trevor Silverstone

Lader, Malcolm. Psychophysiology of Mental Illness. *280 pp.*

● **Silverstone, Trevor** and **Turner, Paul.** Drug Treatment in Psychiatry. *Revised edition. 256 pp.*

Whiteley, J. S. and **Gordon, J.** Group Approaches in Psychiatry. *240 pp.*